JOHN ZANOFF III

A FRAMEWORK FOR COMPLEX SYSTEM DEVELOPMENT

A FRAMEWORK FOR COMPLEX SYSTEM DEVELOPMENT

Paul B. Adamsen II

CRC Press
Boca Raton London New York Washington, D.C.

Library of Congress Cataloging-in-Publication Data

Adamsen, Paul B.
A complex system design and management framework / Paul B. Adamsen, II.
p. cm.
Includes bibliographical references and index.
ISBN 0-8493-2296-0 (alk. paper)
1. Systems engineering.
2. Industrial management. I. Title.
TA168 .A28 2000
620'.001'171—dc21 99-086803
CIP

Direct all inquiries to CRC Press LLC, 2000 N.W. Corporate Blvd., Boca Raton, Florida 33431.

International Standard Book Number 0-8493-2296-0
Library of Congress Card Number 99-086803
Printed in the United States of America 1 2 3 4 5 6 7 8 9 0
Printed on acid-free paper

Abstract

This book outlines a structured framework for complex system design and management. There have been and continue to be many efforts focused on defining the elusive generic System Engineering process. It is suggested that one reason why industry, government, and academic efforts have had limited success in defining a generalized process applicable to many contexts is that the time and logical domains have not been explicitly identified and characterized in distinction. When the logical view is combined with the chronological view, the resulting process often becomes application specific. When these are characterized in distinction, the overall framework is preserved. This book develops a generalized process that maintains this distinction and is thus applicable to many contexts.

The design and management of complex systems involves the execution of technical activities together with managerial activities. Because of the organic connection between these two sets of activities, they must be integrated in order to maximize the potential for success. This integration requires a clear definition of what the system development process is in terms of the technical activities and how they logically interact. In this book, this logical interaction has been defined and is called "control logic." This "control logic" is then used to develop the logical connections and interactions between the managerial and technical activities.

Preface

Several years ago, the author became involved in system engineering process development at General Electric Astro Space (now Lockheed Martin) for two compelling reasons. First, he had been leading a number of advanced spacecraft design studies for various space physics missions and was becoming increasingly frustrated at the lack of order in terms of the flow of activity and information. The work was getting done thanks to excellent subsystem engineers, but there was an appalling lack of order, even chaos to some degree. The author began to see the need to develop a more organized approach to complex system development. His opportunity came in the autumn of 1993 when Astro experienced an unprecedented string of spacecraft failures.

On Saturday, 21 August 1993, contact was lost with the Astro-built Mars Observer spacecraft, just three days before it was to enter orbit around the planet. To the author's knowledge, after intensive investigation, there has been no definitive determination as to the cause of failure. On that same Saturday, NOAA-13, a TIROS weather satellite launched just 12 days prior, experienced a total system failure — most likely the result of an oversize screw that eventually caused the entire electrical power system to fail. About 45 days after that, on 5 October 1993, there was a malfunction during the launch of the Landsat 6 satellite that caused the spacecraft to plunge into the ocean.

In the midst of these failures, Astro was competing for a major low earth orbit spacecraft contract. It was in this context that the opportunity came for the author to join that engineering team for the purpose of developing a sound system engineering approach for the program. He was tasked to develop a structured approach that avoided standard "boiler plate" and reflected how the system would actually be developed in the real world. That was exactly what he wanted to do as a personal goal and professional objective — the second compelling reason he became involved in system engineering process development.

After several months of research, trial-and-error, and prayer, the author developed a new system engineering process that was summarized in the paper, "A New Look at the System Engineering Process — A Detailed Algorithm."[1] That process became the basis for the system engineering

[1] Adamsen, Paul B. Jr., A New Look at the System Engineering Process — A Detailed Algorithm, "Systems Engineering in the Global Market Place," *Proceedings of the Fifth Annual Symposium NCOSE*, Vol. 1, July 22-26, 1995, St. Louis, MO.

training course at Astro, which was taught to several hundred junior and senior engineers. It became the starting point for the author's thesis at MIT, and the seed from which this present work has grown.

This book is intended to provide a *framework* for the design and management of complex systems. It is a generalized *framework*, not an exhaustive exposition. The goal has been to distill the essential aspects of system design into a logical process that accurately reflects what should actually occur on a well-organized development program. This book is relatively brief and succinct, which will hopefully extend its usefulness to busy managers, engineers, and students.

Who should read this book?

- System Engineering Managers
- System Engineers
- Engineers involved in complex system development
- Program Managers
- Senior Managers
- Government Procurement Managers
- Customers
- Proposal Managers
- Engineering Educators and Students
- Research and Development Managers

Acknowledgments

I would like to express thanks to the various professors, administrators, and staff of the Massachusetts Institute of Technology (MIT) System Design and Management (SDM) program for their dedication to the students and their hard work that made the program an enjoyable one. In particular, I would like to mention Dean Thomas L. Magnanti, co-founder of the SDM program, for his example of one who has accomplished much in this life and yet maintains a posture of genuine humility; Prof. Steven D. Eppinger, my thesis advisor, for his helpful comments and encouragement; Prof. John R. Williams for his enthusiasm and encouragement; and Dr. James M. Lyneis for his excellent course on System Dynamics that reshaped much of my thinking, and for his review of Chapter 4.

I would like to thank the following for their reviews of some or all of the manuscript in its various stages of development: Mr. Charles Benet for his review of the ADACS examples, Dr. Madhav S. Phadke for his review of the appendix dealing with Robust Design and QFD, Mr. Louis C. Dollive, Mr. Robert M. Kubow (my father-in-law), Mr. Glenn Davis, Mr. David J. Bean and Mr. John Petheram.

I would like to thank Mr. Henry J. Driesse, Mr. Frank Sweeney, Mr. Alan S. Kaufman, Mr. George Scherer, Mr. John R. Zaccaria, and Mr. Charles L. Kohler for their support at ITT. I would like to thank Mr. Mark Crawley, Mr. John Petheram, Mr. Paul Shattuck, and Mr. Richard Kocinski for their support and encouragement during my employment at Lockheed Martin. Thanks also to Mr. Michael Menzel, who first suggested to me that functional decomposition is dependent upon an assumed concept, and to Mr. Paul Gillet for his input regarding the verification activity.

I would like to express thanks to my church family at Trinity Baptist Church: Pastor Albert N. Martin for his godly example that I long to imitate; Pastor Barton Carlson for his friendship and godly example; Pastors Jeff Smith, Frank Barker, and Lamar Martin for their faithfulness and steadfastness; Miss Elaine Hiller for her many prayers for me and my family; and to the many other members who have upheld us in their prayers.

I would also like to thank my family: Mr. and Mrs. Paul B. Adamsen, Sr., my mom and dad, for their prayers, encouragement, and love; Mr. and Mrs. Robert M. Kubow, my mother- and father-in-law, for their many acts of kindness and generosity to my children, my wife, and to me; my children

— Paul, David, and Lauren — for their prayers, patience, and love; and my beloved wife, Karen, for her prayers, support, patience, friendship, and love.

Finally, I would like to express thanks to my Lord Jesus Christ, who, in answer to my prayers and the prayers of many of God's people, has given me a measure of understanding in the area of complex system development. *Soli Deo Gloria.**

* The views expressed are those of the author, and do not necessarily reflect the views of the staff or management of CRC Press LLC.

Dedication

This book is dedicated to my beloved wife and best friend, Karen
and to my children
Paul, David, and Lauren

Contents

Tables and Figures

Tables

Figures

Acronym List

ADACS	attitude determination and control system
AKM	apogee kick motor
ASME	American Society of Mechanical Engineers
Arcmin	arcminute
Arcsec	arcsecond
CAD	computer aided design
CAE	computer aided engineering
CAM	computer aided manufacturing
CCB	configuration control board
C&DH	command and data handling system
CGRO	Compton Gamma Ray Observatory
CMND	command
DP	design parameter
DOD	Department of Defense
DSM	design structure matrix
DSP	defense support program
DSMC	Defense Systems Management College
EIA/IS	Electronic Industries Alliance/Interim Standard
EMC	electro magnetic compatibility
EMI	electro magnetic impulse
EPS	electrical power system
EQFD	enhanced quality function deployment
ERS	earth research satellite
ESAT	example satellite
ETM	engineering test model
FAT	fabrication, assembly, and test
FDIR	failure detection isolation and recovery
FMECA	failure modes effects and criticality analysis
FR	functional requirement
FSS	fine sun sensor

GEOSAT	Geodetic satellite
H/W	hardware
HST	Hubble Space Telescope
Hz	hertz
IEEE	Institute of Electrical and Electronics Engineers
INCOSE	International Council on Systems Engineering
ICD	interface control document
I&T	integration and test
I/F	interface
I/O	input/output
IMU	inertial measurement unit
Kbps	kilobit per second
Km	kilometer
MIT	Massachusetts Institute of Technology
NASA	National Aeronautics and Space Administration
NOCSE	National Council on Systems Engineering
NOAA	National Oceanic and Atmospheric Administration
OBC	on-board computer
OP'S	operations
PDP	product development process
P/L	payload
PRS	propulsion system
QFD	quality function deployment
R&D	research and development
RD	requirements development
RF	radio frequency
RQMT	requirement
RW	rework
RWA	reaction wheel assembly
S/C	spacecraft
SDF	system development framework
SMS	structure and mechanism system
SS	subsystem
S/W	software
TBD	to be determined

TBR	to be reviewed
TBS	to be supplied
TCS	thermal control system
TDRS	tracking and data relay satellite
TDRSS	tracking and data relay satellite system (includes ground terminals)
TDW	tailored document worksheet
TIROS	television infrared observation satellite
TLM	telemetry
TPM	technical performance measure
TSE	total system elements

chapter 1

Introduction

This book deals with the problem of the design and management of *complex* systems. Dommasch and Laudeman comment, "It is well to remember that any fool can design a complex system to do a simple job, but it is something else again to design a simple system to perform a complex task."[2]

In order to develop the simplest solution to a complex problem, the problem must be well understood. Thus, a central concern in the development of complex systems is understanding. This involves the ability to understand the customer's needs that are to be addressed by the system; to understand the context in which the system will function; to understand the solution to the problem as it is being developed; and finally to help the customer understand his/her problem and its proposed solution. As systems continue to become more complex, the problem of understanding becomes more acute. The author's background is spacecraft systems. Like many complex systems, early spacecraft were relatively simple. However, especially after the advent of the computer, spacecraft have become increasingly complex.

I. Is a Structured Approach Needed?

While acknowledging that many companies do employ a structured approach to the development of their systems, many do not. Therefore, the following are offered as suggestions as to why a structured approach to complex system development is increasingly necessary.

Increasing Complexity — First, as just mentioned, systems are becoming increasingly complex. It used to be that the success of a new system development activity often depended upon the competence of the system engineer who knew everything about it. However, the days of those kinds of efforts being successful are numbered. Systems are becoming so complex that it is increasingly difficult to find any one person who is able to "keep it all in his head."

[2] Dommasch, Daniel O. and Charles W. Laudeman, *Principles Underlying Systems Engineering*, New York: Pitman Publishing, 1962, p. 393.

Personnel Longevity — A second reason has to do with the longevity of people. The explosion of complexity, in large part, has taken place within the lifetime of the last generation. As those people who grew up with the industries now producing complex systems retire or otherwise leave the industries, who is left to "pick up the ball"? New people, who did not have the opportunity to learn the systems when they were relatively simple. These new people are left with the task of "picking up the ball," not at the size of a snowflake, but after a system has snowballed to often avalanche-creating proportions. How are these people going to be able to do this effectively?

Workforce Discontinuities — Third, long careers in the same organization are becoming increasingly rare. Companies continue to merge, divest, reorganize, acquire, consolidate, relocate, globalize, centralize, and so on. Likewise, employees are becoming increasingly mobile, moving from one company to another every several years. This causes an instability and discontinuity in the workforce and therefore also on the development programs of which they are a part.

Intense Competition — Fourth, intense competition and limited financial resources are putting enormous pressure on development programs to reduce cost and cycle time. In order to remain competitive, these programs must be executed in an increasingly efficient manner.

This is certainly not an exhaustive list, but these things do identify a need to develop effective ways of dealing with the dynamics of a changing environment. Clearly there are many elements to an effective strategy of dealing with these and other pertinent issues. A central element of an effective strategy must include effective structuring of complex system development activities.

II. Technical and Managerial — Integration is Essential

The design and management of complex systems involves the execution of technical activities (e.g., requirements development, design, analysis, integration, verification, trade analyses, etc.) together with managerial activities (e.g., configuration management, risk management, etc.). Because of the necessary connection between them, these two sets of activities must be integrated. But how should this be done? The approach discussed herein involves clearly defining what the system development process is in terms of the technical activities, and then using it to develop the logical connection between the managerial and technical activities.

III. Motivation

Why study the topic of complex system design and management? The author concurs wholeheartedly with Wymore, who comments:

> Every author has several motivations for writing, and authors of technical books always have, as one motivation, the personal need to understand; that is, they write because they want to learn, or to understand a phenomenon, or to think through a set of ideas."[3]

Most anyone who has been involved in the development of a complex system has been struck by the amount of chaos that can develop so quickly. Requirements are not communicated to the right people. Sufficient resources are not available when they are needed. The customer does not know exactly what he wants or needs, thereby causing an instability in the top-level requirements. Heritage hardware and software bid in the proposal do not support the interfaces of the new system or perform to the levels required. Back-of-the-envelope analyses and engineering judgment do not agree with current, more detailed analyses, thus invalidating previous assumptions. The list could go on and on. It is this problem of controlling the chaos in complex system development that inspired the author's fascination.

IV. Objectives

A central objective of this book is to develop a generalized approach to system development, such that the chaos which inevitably ensues can be controlled. More specifically, key objectives include:

- Developing an overarching generalized process that integrates all technical and managerial activities, that is applicable to all development phases from conceptual to detail design, and that is applicable to all levels of the design from top-level system to the lowest levels of the system hierarchy;
- Defining the System Development Framework (SDF) in sufficient detail so as to enable its implementation in the real world;
- Deriving a set of principles that serve to guide the person or teams involved in complex system development;
- Laying the foundation necessary to accurately model the System Development process;
- Clarifying where in the process iteration occurs and why; and
- Clarifying how functional decomposition is performed and its necessary relationship to implementation. This involves understanding the relationship between the "what" provided in specifications and the "how" developed in the implementation.

[3] Wymore, Wayne A., *A Mathematical Theory of Systems Engineering — The Elements*, New York: John Wiley and Sons, 1967, p. v.

V. Key Questions

When setting up a program, the management team faces several important decisions that can have a significant impact on the performance of the contract. Some of these include:

- How should information flow and who is responsible for which interfaces?
- What are the technical activities and how should they be ordered?
- How should the managerial aspects of system development be structured?
- How should the technical and managerial activities be coupled?

These questions are addressed in this book.

VI. "System" Defined in the Literature

Since the main subject of this book has to do with the development of complex systems, it is necessary to discuss how the term is used herein. Hall says, "Systems Engineering is probably not amenable to a clear, sharp, one sentence definition."[4] There are, however, several definitions found in the current literature, and Hall offers his own: "A system is a set of objects with relationships between the objects and between their attributes."[5] Dommasch and Laudeman define "system" as follows:

> A complete system is any complex of equipment, human beings, and interrelating logic designed to perform a given task, regardless of how complex the task may be. Logically, very large or complicated systems are broken into subsystems, to be fitted together like blocks to form the entire or total system.[6]

In his 1991 work, Rechtin defines a system as "a collection of things working together to produce something greater. . . . A system has the further property that it is unbounded — each system is inherently a part of a still larger system."[7] He continues:

1. A system is a complex set of dissimilar elements or parts so connected or related as to form an organic whole.

[4] Hall, Arthur D., *A Methodology For Systems Engineering*, Princeton, NJ: D. Van Nostrand, 1962, p. 4.
[5] *Ibid.*, p. 60.
[6] Dommasch and Laudeman (1962), p. 182.
[7] Rechtin, Eberhardt, *Systems Architecting: Creating and Building Complex Systems*, Englewood Cliffs: Prentice-Hall, 1991, p. 1.

2. The whole is greater in some sense than the sum of the parts, that is, the system has properties beyond those of the parts. Indeed, the purpose of building systems is to gain those properties.[8]

Rechtin with Maier suggest, "A system is a collection of different things which together produce results unachievable by themselves alone. The value added by systems is in the relationships of their elements."[9] Chase[10], Wymore (1967)[11], Ellis and Ludwig[12], and Minds[13] define a system as essentially "anything that performs work on an input in order to generate an output."

VII. Working Definition of "System"

There is overlap between the physical system being developed and the processes used to effect the development. In the context of product development, Krishnan et. al. suggest that the product development process involves the

> . . . transformation of input information about customer needs and market opportunities into output information which corresponds to manufacturable designs. . . . Individual development activities are themselves viewed as information processors, receiving input information from their preceding activities, and transforming this input information into a form suitable for subsequent activities.[14]

This is an excellent summary statement as to how each element of the development process is viewed, as well as how each element of the system itself is viewed in this book. Any complex system composes several subsystems, which, in turn, compose more sub-subsystems. Each of these "systems" receives input, does work on that input, and then generates an output. This is an important point because it suggests that any process applicable to the development of a system must, by this definition, be applicable to the development of its subsystems. This principle lays the foundation for a modular approach to the process definition.

[8] *Ibid.*, p. 28.
[9] Rechtin, Eberhardt and Mark W. Maier, *The Art of Systems Architecting*, Boca Raton, FL: CRC Press, 1997, p. 21.
[10] Chase, Wilton P., *Management of System Engineering*, New York: John Wiley & Sons, 1974, p. 54.
[11] Wymore (1967), p. 21.
[12] Ellis, David O., Fred J. Ludwig, *Systems Philosophy*, Englewood Cliffs: Prentice-Hall, 1962, p. 3.
[13] Minds, Kevin S., "System Engineering The People System," NCOSE 1995, P066.
[14] Krishnan, Viswanathan, Steven D. Eppinger, Daniel E. Whitney, "A Model-Based Framework to Overlap Product Development Activities," *Management Science*, Vol. 43, No. 4, pp. 437-451, April 1997.

For this book, the term "system" will be defined as "any entity within prescribed boundaries that performs work on an input in order to generate an output." Note further that a system's external interfaces consist of those entities or realities that impinge upon the prescribed boundaries. The system exists within a definable physical space and within a specific timeframe.

chapter 2

Literature Search and Rationale for this Book

You can observe a lot just by watchin'.
—Yogi Berra

You can't get where you want to go if you don't know where you are.
—Anonymous

I. Existing and Emerging Standards

While the act of system engineering is as old as man's first efforts to engineer systems, as a defined discipline it is relatively new.[15] Although effort is being expended to change things, in many circles, there remains little consensus on nomenclature, metrics, or the system engineering process itself. However, as will be indicated in this chapter by way of a survey of the literature, there is significant agreement regarding which activities must be performed as part of the system engineering process.

Table 2.1 provides a summary survey of two emerging non-government standards (IEEE 1220-1994, EIA/IS-632)[16] as well as three key military standards (Defense Systems Management College [DSMC] Systems Engineering Management Guide, Mil-Std-499A, and Army Field Manual 770-78).

II. Individual Works

The literature provides many views of the system engineering process. Shinners defines a "closed loop, iterative process."[17] Reinert and Wertz define

[15] For example, Dommasch and Laudeman assert that the term "system engineering" originated with the Manhattan Project. *Principles Underlying Systems Engineering*, p. iii.
[16] See Rechtin and Maier for a critique/comparison of these two standards, pp. 218-219. Note also the Mil-Std-499B was never released and has subsequently evolved to EIA/IS-632.
[17] Shinners, Stanley M., *A Guide to Systems Engineering and Management*, Lexington, MA: D. C. Heath and Company, 1976, pp. 14-17.

Table 2.1 Civilian and Military Systems Engineering Standards

IEEE 1220-1994[a]	EIA/IS-632[b] (Was Mil-Std-499B[c])	DSMC Systems Engineering Management Guide[d]	Mil-Std-499A[e]	Army Field Manual 770-78[f]	Consensus
Requirements Analysis	Requirements Analysis Functional Analysis/ Allocation	Functional Analysis	Mission Requirements Analysis and Functional Analysis	Function Analysis	**"What"**
Synthesis	Synthesis	Synthesis	Allocation and Synthesis	Synthesis	**"How"**
Systems Analysis	Systems Analysis and Control	Evaluation and Decision	Optimization: Effective Engineering Analysis	Evaluation and Decision	**"How Well"**
Functional and Physical Verification	Verification (defined as a feedback)		Production Engineering Analysis		**"Verify"**
Trade Studies and Assessments		Evaluation and Decision	Optimization: Trade Studies	Evaluation and Decision	**"Select"**

[a] IEEE 1220-1994, *Trial-Use Standard for Application and Management of the Systems Engineering Process*, Clause 6, 1995.

[b] EIA/IS 632, Systems Engineering, pp. 7-12.

[c] MIL-STD-499B.

[d] DSMC, *Systems Engineering Management Guide*, 1990, pp. 5-1 to 8-19, Cf. DSMC 1986 *Systems Engineering Management Guide.*

[e] MIL-STD-499A, Engineering Management (USAF), Chapter 10.

[f] Army Field Manual 770-78, April 1979, Figure 2-1.

a "concept exploration flow."[18] Coutinho describes his process as a "systems development and test cycle."[19] Hall delineates a view that captures "the open systems nature of the systems engineering process as it exchanges energy, information, and materials with the environment."[20] Blanchard and Fabrycky emphasize a "sequential and iterative methodology to reach cost-effective solutions to design alternatives." They continue, "Systems engineering is directly concerned with the transition from requirements identification to a fully defined system configuration ready for production and ultimate cus-

[18] Reinert, Richard P. and James R. Wertz, *Space Mission Analysis and Design*, Eds. Wiley J. Larson and James R. Wertz, published jointly by Microcosm, Inc., Torrance, CA and Kluwer Academic Publishers, The Netherlands, 1992, p. 20.

[19] John de S. Coutinho, *Advanced Systems Development Management*, New York: John Wiley & Sons, 1977, pp. 35-51.

[20] Hall, pp. 138-139.

tomer use."[21] Chase identifies what he calls "the irreducible gross functional steps which must be followed [in any system development activity]."[22] Wymore has developed a list of "archetypal" questions that are reflective of a sound system engineering process: What is the system supposed to do? What is available to build the system? How well must the system perform? What are the cost/performance trades? How can the system be verified?[23] Each of these are summarized in Table 2.2.

As Tables 2.1 and 2.2 indicate, there is general consensus among the various system engineering process standards in terms of the technical activities that must be performed.[24] Each of the general activities identified in each of the sources surveyed may be categorized into one of the following five broad categories: "what," "how," "how well," "verify," or "select" activity. It is asserted that *any* technical development activity can be categorized under one of these categories. This is the basic organizing concept of the System Development Framework (SDF) derived in this book and is illustrated in Figure 2.1.

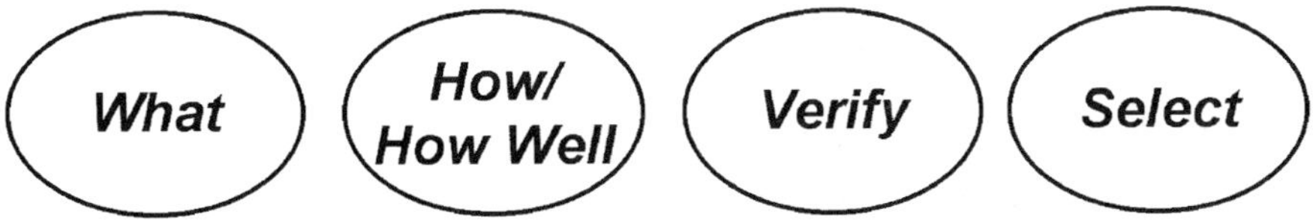

Figure 2.1 System Development Framework (SDF) Organizing Concept.

III. The Basic Building Block

There is a logical flow to the sequencing of these activities. There must, first of all, be some input, however high or low its fidelity. This input must be analyzed to determine "what" the system must do. This activity is called Requirements Development. The next activity — Synthesis — generates alternatives describing "how" the "what" might be implemented. Synthesis also determines "how well" each alternative performs, as well as verifying that the design meets the objectives. If more than one alternative emerges, the best is chosen through a "selection" process. To reiterate, Figure 2.1 illustrates this top-level flow of activity which is the SDF Organizing Concept. It is used to organize the various engineering activities and subprocesses into an overall integrated process.

[21] Blanchard, Benjamin S. and Wolter J. Fabrycky, *Systems Engineering And Analysis*, Englewood Cliffs, NJ: Prentice-Hall, 1981, pp. 236-240.
[22] Chase, pp. 7-11.
[23] Wymore, Wayne A., *Model-Based Systems Engineering*, Boca Raton, FL: CRC Press, 1993, p. 7.
[24] Cf. White, Michelle M., James A. Lacy, and Edgar A. O'Hair, Refinement of the Requirements Definition (RD) Concept in a System Development: Development of the RD Areas, "Systems Engineering Practices and Tools," *Proceedings Sixth Annual Symposium INCOSE*, Vol. 1, July 7-11, 1996, Boston, MA, pp. 749-756.

Table 2.2 Individual Works

Shinners	Reinert and Wertz	Coutinho	Hall	Blanchard and Fabrycky	Chase	Wymore	Consensus
Mission and Requirements Analysis and Functional Analysis	Define Preliminary Requirements	Operational Requirements	Problem Defining	System Functional Analysis	Mission Objectives, System Performance Requirements, Operations and Support Requirements	"What" is the System Supposed to Do?	**"What"**
Synthesis of the System	ID Concepts, Define Mission Architecture	System Design	Synthesis and Design	Synthesis and Design	System Design and Synthesis	What is Available to Build the System?	**"How"**
Effectiveness Analysis Simulations	Evaluate Concepts	Systems Analysis	Analysis	Optimization and Analysis	System Analysis	"How Well" must the System Perform?	**"How Well"**
Test of System	Generate Measures of Effectiveness	Prototype Design, Fab and Test Plan, Acceptance	Test	Prototype, Test and Evaluation	Evaluation of System Design Adequacy	How can the System be Verified?	**"Verify"**
Tradeoff Analysis	System Level Trades		Decision Making		Select Alternative System Design Approaches, Perform Trade-Off Studies	What are the Cost/ Performance Trade-Offs?	**"Select"**

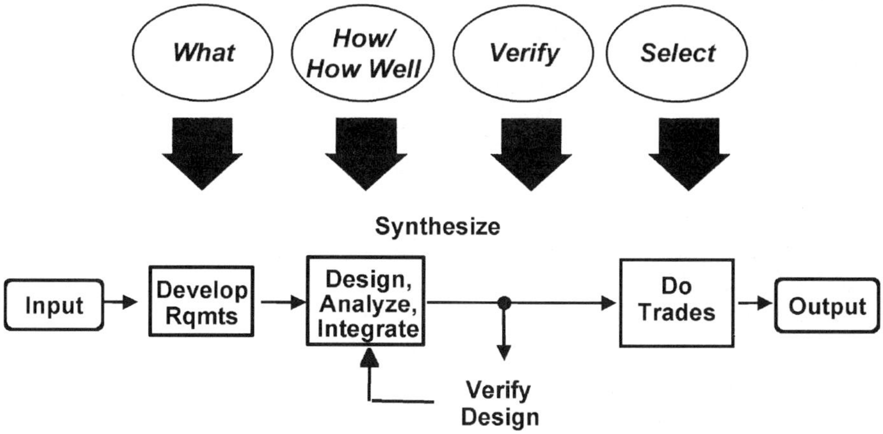

Figure 2.2 The SDF Basic Building Block

Figure 2.2 represents this sequence as the basic building block of the SDF. It represents one "module" of activity. It is applied to each system, subsystem, sub-subsystem, etc. of the program hierarchy. A detailed lower level decomposition of this basic building block is developed in Chapter 5.

IV. Unique Features of this Book

In the above discussion, the fundamental building block of the SDF has been developed, based upon a consensus derived from the system engineering literature. At this point one may ask, "Why another book on the system engineering process?" There are several unique features in this present work.

A. Time and Logical Domains

The time and logical domains have been identified and characterized in distinction, thus enabling the SDF's application to many contexts. There have been and continue to be many efforts focused on defining the elusive generic System Engineering Process. It is suggested that one reason why industry, government, and academic efforts have had limited success in defining a generalized process applicable to many contexts, is that the time and logical domains (discussed in detail in Chapter 3) have not been explicitly identified and characterized in distinction. When the Logical Domain view is combined with the Time Domain view, the resulting process often becomes application specific. When these are characterized in distinction, the overall framework can be preserved. This book develops a generalized process that maintains this distinction and is thus applicable to many contexts.

B. Tier Connectivity

Many system engineering processes are defined with reference to one tier of activity. They do usually acknowledge multiple tiers (i.e., system subsystem, sub-subsystem). However, in general, they do not precisely describe how the various tiers are coupled in terms of logical and consistent flow-down and feedback paths. This book clearly defines the coupling between tiers, which has implications regarding information flow, roles and responsibilities, change impact analyses, risk management, etc.

C. Modularity

The SDF defined in this book is truly modular. Through prudent partitioning of the system into focused modules, the SDF is tailored to meet the demands of individual development programs.

D. Coupling of Technical and Managerial Activities

The design and management of complex systems involves the execution of technical activities together with managerial activities. Because of the organic connection between these two sets of activities, they must be integrated in order to maximize the potential for success. This integration requires a clear definition of what the system development process is in terms of the technical activities and how they logically interact. In this book, the "control logic" (see Chapter 5) provided by the SDF is used to develop the logical connection between the managerial and technical activities.

E. Clear Presentation of Functional Decomposition

In some circles, there is significant confusion regarding system development by functional analysis and decomposition. This book attempts to provide a clear and logical approach to this important activity.

F. Explicit Inclusion of the Rework Cycle

Within the discipline of System Dynamics, the rework cycle was developed some years ago and it has been adapted into the overall SDF. This is essential for an accurate understanding of real-world system development dynamics and accurate modeling of any development activity.

G. Explicitly Defined Generalized Outputs

As one 40-year veteran system engineer put it, "Many system engineering processes have been defined, but few remember the main point — output." Therefore, an indication of what each activity must produce in terms of generalized output has been provided. That output then becomes the input to subsequent activities.

chapter 3

System Development Framework (SDF) Overview

If you don't know where you're going, you'll end up somewhere else.
—Yogi Berra

In Chapter 2, the basic building block of the SDF was developed, based upon the collection of activities that is commonly associated with the System Engineering Process. These activities were derived from the consensus that emerged through a survey of the literature. It was suggested that there was little consensus, however, in terms of the arrangement of those activities into a consistent and coherent process. In this chapter, a top-level framework in which each of the identified activities can be arranged is developed.

In many industries, fierce competition is forcing cost and schedule resources to decrease substantially. In some industries, new programs are moving from financially safe cost-plus contracts to more risky fixed price contracts.[25] This puts much more pressure on the contractor to contain costs so that overruns that detract from profit can be minimized or avoided altogether. In order to increase the probability of earning a reasonable profit, it is essential that the bid reflect an accurate assessment of the scope of effort involved. In order to develop an accurate assessment of scope, an accurate understanding of the tasks involved and the resources required to perform those tasks is critical. Therefore, the more accurately the process for developing the system is defined, the more accurately the scope of the necessary effort can be determined.

[25] Cost-plus contracts generally assure the contractor that all costs incurred during the performance of the contract will be reimbursed by the customer. This type of contract is usually implemented where there is significant risk in the development of a new system or technology. In this type of contract, the customer agrees to bear some or all of the financial risk. Fixed price contracts, on the other hand, require that the system or product be delivered for the price agreed to by the contractor and customer at the time of the contract award regardless of the actual costs incurred by the contractor.

I. Two Views Needed for an Accurate Model

The system engineering process must be defined with respect to two scales of time — the macro-scale and the micro-scale. In terms of the macro-scale, program development over large increments of time or program phases spanning from initial studies to the operational phase of the program is considered. This view of the program is called the Time Domain view. On the micro-scale, the concern is with information flow and energy expenditure at any instant in time. This view has been called the Logical Domain view. These two domains are more fully discussed below, but first the rationale for this approach is considered.

A. Rationale

Why emphasize this distinction between the macro and micro time scales? A central goal of this book is to define a process that accurately reflects what ought to occur on a well-ordered system development program in the real world. And why is this important? There are several reasons:

- **Cost Containment** — Provide a basis for accurate modeling of the program to facilitate more accurate estimates to complete, and for more accurate change impact analyses. This will serve to maximize the probability that the program will be completed within its cost and schedule constraints.
- **Scope Assessment** — Provide a basis for accurately assessing the scope of a development program which, in turn, facilitates more accurate bids for new contracts with more solid bases of estimate.
- **Metrics Development** — Provide a basis for accurate measuring of the state of the program as the development progresses.
- **Resource Planning** — Provide a basis for determining when specific resources will be required.

So, the first reason for this distinction is to provide an accurate model of what actually occurs in the real world of system development programs. A second goal of this book is to define a system development framework that can be applied to a wide variety of contexts. It is suggested that when the Time and Logical Domains are not explicitly identified and characterized in distinction, much difficulty arises in terms of defining a coherent and consistent development process that can be applied in a wide variety of contexts.

B. An Illustration

Consider the activity commonly called Verification. Verification is an activity that often consumes a significant percentage of overall program resources for its execution. Most of those resources are generally expended after the design work has been completed. In fact, some system engineering processes define this activity as occurring after the last major design review. However, is that an accurate or even desirable depiction of when this critical activity should take place — at the end of the design phase?

In order to minimize risk to the program, Verification issues must be considered at the earliest stages of development. Where will the system be tested? What facilities will be available? What interfaces must be satisfied? How will the system be transported? What special equipment will be needed? If these issues are considered during the early stages of design it will likely be possible to minimize the cost and schedule needed to perform the Verification activity.

Given that Verification must be considered throughout the system development, how should this be depicted in the System Engineering Process? The same question could be asked about any of the activities identified in Chapter 2. What about Requirements Development — does that end with the first major review? What about functional analysis, design, allocation, design integration, all the various analyses that must be performed, trade studies, etc.? Do all these occur in a strict serial fashion? Once these have been performed at some level, are they completed for the duration of the development? What about other levels in the hierarchy — is Verification performed only at the system level? The obvious answer to each of these questions is "No." Most of the activities of the system engineering process occur in parallel within the same hierarchical level of the development.[26] They also occur across the development timeline within the design activities of subsystems and sub-subsystems and so forth. But how should this be depicted as a logical process?

It is suggested that, in order to accurately capture what actually occurs in the real world of complex system development, the system engineering process must be defined in two domains — the macro-scale, or Time Domain, and the micro-scale, or Logical Domain. In the former, the design evolution that should occur over large increments of time is planned (design phases, manufacturing, integration and test, etc.). In the latter, the logical flow of the technical activities (Requirements Development, Synthesis, and Trade Studies) that should be occurring at any instant in time is defined.

Now, returning to the example, how should each of the technical activities be represented in the overall development process? It has been pointed out that each activity is generally performed at some level of intensity and for each system element during the entire development phase. Requirements Development, Synthesis, and Trade Study activities are all performed at different levels of intensity across the program timeline. These activities are generally performed in parallel, not in series. They often

[26] Brooks speaks of the "classical sequential or waterfall model," p. 265. He then goes on to offer a critique, "The basic fallacy of the waterfall model is that it assumes a project goes through the process once, that the architecture is excellent and easy to use, the implementation design is sound, and the realization is fixable as testing proceeds," p. 266.

overlap as one activity feeds the next with data that is not necessarily complete. It is also generally agreed that these activities are iterated over the course of the development. So, how should this be depicted? If it is agreed that each of these activities generally occurs on some element of the system at some level in the hierarchy at some level of intensity, then each activity must be depicted as occurring at each point on the program timeline. It is not realistic to depict each of the technical activities as occurring in strict serial fashion along the program timeline. Therefore, it is suggested that the concept of two domains be employed because this facilitates an accurate model of what actually occurs in the real world of complex system development.

II. Time and Logical Domain Views Provide a Full Program Description

As mentioned above, the macro-scale, or Time Domain, considers the System Development activity as viewed across the entire life cycle of the program. It is concerned with how inputs and outputs evolve over time as the system design matures. The micro-scale or Logical Domain, on the other hand, deals with what occurs within small increments of time. Imagine instantaneous "snapshots" along the macro-timeline, or time continuum, where there are an infinite number of "logical planes." Each "snapshot," or plane normal to the macro-timeline, represents an infinitesimally small slice of time. Each slice, or plane, reveals the instantaneous logical sequencing of activity occurring at that time. Figure 3.1 illustrates the Time Domain view of the program and the instantaneous "snapshots" that are provided by the Logical Domain view. It further depicts how these two views combine to fully define the program.

A. Time Domain Focus: Inputs and Outputs

The Time Domain view characterizes how the design evolves chronologically. The essential "value-added" element of this view is a clear definition of inputs and outputs as they are developed over time. The focus of this view is not the activities being performed, but rather the outputs generated by those activities. It is the outputs that change significantly from phase to phase. The definition of these outputs during the planning process becomes one of the primary bases by which the program scope is assessed.

Note here that the output of a previous activity becomes the input to the subsequent activity. As Figure 3.1 indicates, there are two distinct sets of data that are output: requirements and design. Notice that the requirement outputs lead the design outputs. The logic is that the requirements at any given level of the hierarchy drive the design being generated at the same level. This is discussed in more detail in Chapter 5.

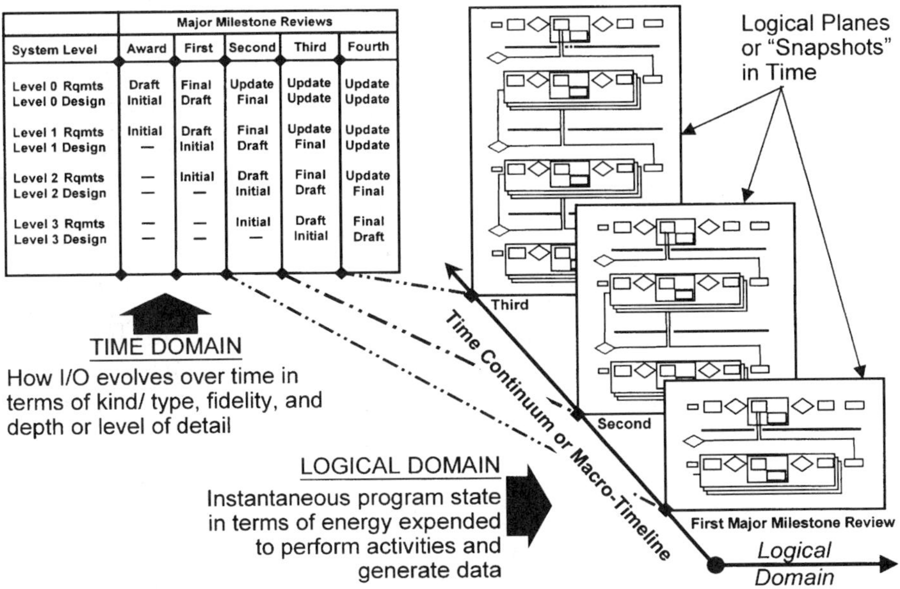

Figure 3.1 Time and Logical Domain Coupling.

B. Logical Domain Focus: Energy Expenditure

The Logical Domain view provides an instantaneous snapshot in time of the program state, revealing which activities are being performed at each hierarchical level, what the sequencing of those activities is, and how much energy is being applied in the performance of each activity.

Energy is expended in each activity (i.e., requirements development, synthesis, and trades) until the desired output at the necessary fidelity is generated. This "energy" refers to the manpower and other resources needed to generate the required output. It is the scope of the effort necessary. Thus the more accurate the estimates for outputs required and the inputs needed to generate the outputs, the more accurate the manpower and other resource estimates will be.

In any well-planned complex development program there are major milestones where a certain level of design definition is planned to be achieved. Figure 3.1 illustrates three such milestones. The first major milestone is concerned with defining the top-level system architecture. The second focuses on defining the architectures of the various subsystems. The third and subsequent major milestones generate the designs of the lower level elements. Each plane reveals how many levels of the hierarchy are involved and how many subsystems are at each level, what the logical connections are within and between each of the levels, and the energy level applied to each activity.

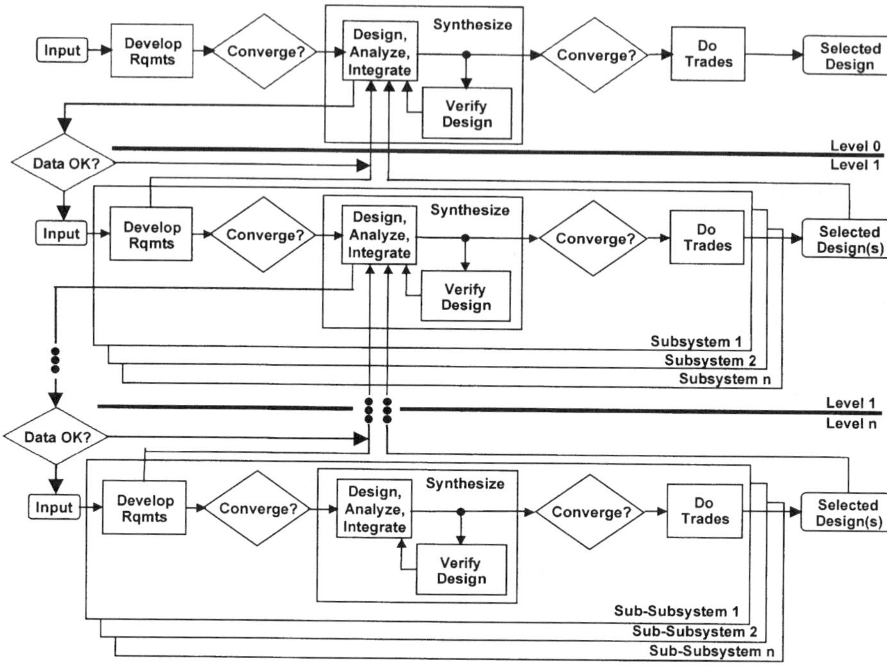

Figure 3.2 The SDF Logical View.

III. The SDF in the Logical Domain

Figure 3.2 enhances Figure 2.2 of Chapter 2 by including decision points for convergence, adding tiers to the program, and indicating connectivity between the tiers. It illustrates that there is control logic coupling the various modules of activity between and within the tiers of the system hierarchy. Several important points are implied by Figure 3.2.

A. Control Logic

One of the key elements of the SDF is the clear delineation of the data flow-down and feedback paths that connect same-tier and adjacent-tier activities. This "control logic" defines the paths and the gates through which information flows within the overall hierarchy. This is developed in detail in Chapter 5, where Figure 3.2 is decomposed to the next level.

B. Hierarchy

The Moses of the Bible implemented a hierarchy of leadership in his governing the nation of Israel (Exodus 18:13-27). Hierarchy is an essential element of organizing any complex system. The SDF building block, mentioned above, defines the basic activities and the logical sequencing of those activities for each tier in the program hierarchy.

At any instant in time it is likely that each activity of the SDF is being performed at some level of intensity and at some level in the system hierarchy. The level of intensity applied to each activity is dependent upon a whole array of variables: stability of the input requirements, level of complexity of the system, whether the system is precedented or not, where on the program timeline the development effort is occurring, etc.

C. Modularity

The SDF is modular and therefore tailorable. Tailoring is accomplished by partitioning the program hierarchy appropriately, adding tiers as necessary, and by adding same-tier elements as needed.

D. Closed Loop

The SDF is closed loop, with information flowing down from the Design, Analyze, and Integrate activity to the next tier, and with data flowing back up from the lower tier into the Design and Integrate activity of the tier above. This is an important point because it is this organization that can preclude design teams from spending resources and increasing costs by designing in capabilities that are not required. Conversely, it can also be used to ensure design teams are responding to all design requirements, thus eliminating expensive redesign needed to include functionality that was missed.

E. Traceability

The SDF provides a structure for program requirements databases. This logical flow of activity ought to reflect the manner in which requirements flow up and down throughout the system hierarchy. Therefore, this same structure should be used in designing the requirements traceability system, which includes not only requirements flow, but also verification method and the stage in the system build-up at which the verification will be performed. It also provides the basis for change impact analyses, sensitivity analyses, cycle-time reduction analyses, etc.[27]

F. Comprehensiveness

It is important to note that each activity of which the SDF is composed considers not only the development of the deliverable product itself, but also all of the associated hardware, software, procedures, processes, etc. needed to produce, integrate, test, deploy, operate, support, and dispose of the system. The effort applied to the development of support elements is commensurate with program need.

[27] Cf. Adamsen (1995)

G. Convergence

With regard to the flow of the process, the key criterion in moving from Requirements Development (RD) to Synthesis is convergence of the RD activity. Requirements convergence has to do with defining a set of requirements that are stable enough to proceed to the Synthesis activity with acceptable probability that an adequate solution can be generated. The requirements set will never reach absolute perfection. *The issue is whether or not the risk associated with proceeding to the next activity is acceptable.*

An example of non-convergence occurred on a space program in which the author was involved. The spacecraft was required to communicate with a relay satellite during a time when the two prescribed orbits precluded such communication. Because this was an impossible requirements set to satisfy, the requirements activity could not converge. The requirements set needed to be changed in order to enable the Requirements Development activity to converge upon a solution. Another example is the case in which a function cannot be performed without input from another function. In such a situation, the input and output requirements of the functions must be coordinated. If this is not possible, the Requirements activity can not converge.

The question of convergence relative to the Synthesis activity is similar to that of the Requirements activity. Examples of non-convergence might include the unavailability of certain technologies required to satisfy a particular requirements set. The requirements might be perfectly stable and consistent, but they are not implementable — at least not at a reasonable cost or schedule. A feedback path to the input source is provided in the process when convergence is not achieved.[28]

To summarize, convergence occurs when the output data has achieved a level of acceptable risk in terms of the probability of success that the subsequent activity will be able to reach convergence with that data.[29]

H. Risk

Risk management is an important tool by which to manage the development of any complex system. The primary difficulty arises in assessing it accurately and in a meaningful way. Nevertheless, a risk assessment of the output data ought to be an essential criterion in making the determination whether or not to proceed to the subsequent activity. The risk assessment should focus on determining if the requirement and design output are of such sufficient fidelity that the downstream activities can commence with an acceptable probability of success.

[28] In order to avoid clutter in Figure 3.2, the feedback paths to the input source are not shown. They are, however, shown explicitly in Figure 5.42 of Chapter 5.

[29] A key issue here is accurately quantifying the risk to the program if the subsequent activity proceeds with the data input to it.

System Tier	Award	First	Second	●●●	"N"
Level 0 Rqmts	Draft	Final	Update	●●●	Update
Level 0 Design	Initial	Draft	Final		Update
Level 1 Rqmts	Initial	Draft	Final	●●●	Update
Level 1 Design	—	Initial	Draft		Update
Level 2 Rqmts	—	Initial	Draft	●●●	Update
Level 2 Design	—	—	Initial		Final
Level "n" Rqmts	—	—	Initial	●●●	Final
Level "n" Design	—	—	—		Draft

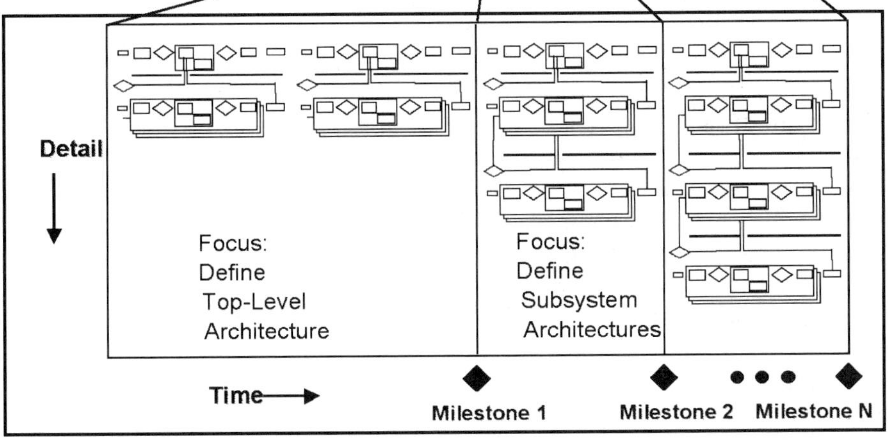

Figure 3.3 The SDF in the Time Domain.

IV. The SDF in the Time Domain

Having described how the SDF functions in the Logical Domain, the question of how it functions in the Time Domain is now addressed. Figure 3.3 depicts the SDF in the Time Domain. During the early stages of the development, the primary focus of activity centers on defining the top-level system architecture. The figure illustrates several top-level architecture candidates being evaluated during the first phase of activity. The conclusion of that phase of development culminates in the selection of the baseline architecture. After that point, the focus of activity moves to development of the major subsystems at the next level down in the hierarchy. This introduces the concept of "incremental solidification."

A. Incremental Solidification

In order to proceed along the program timeline with minimal risk, it is necessary to "incrementally solidify" key requirements by program milestone. It is highly desirable to determine which requirements are needed from the customer (or other entity) and when, in order to maintain progress with minimal risk. One goal is to include these critical need dates in the contract so that if there is delay, a cost and schedule scope change can be effectively negotiated. It is also apparent from the figure that an instability in upper-tier requirements or design results in instability at each dependent tier below it. Thus, unstable requirements or design at an upper level induce risk into lower dependent levels.

B. Risk Tolerance Defines Scope

Notice in Figure 3.3 that only one level of activity below the top-level has been shown during the first phase. This is intended to illustrate that, at early stages of development, lower levels of design and analysis are only performed in order to support the development of the top-level architecture. In other words, the degree of lower-level analysis necessary is determined by the amount of risk the program is willing to carry. If there is little uncertainty that the lower level design can perform as required, then there is little reason to perform detail design and analysis at that point on the timeline. Conversely, if there is great uncertainty that a key lower-level design will be able to perform as required, then significant lower level design and analysis may be necessary to lower the program risk to an acceptable level.

To illustrate this point, consider the conceptual design of a spacecraft constellation in which direct communication between satellites within the constellation is necessary. The method of communication is a key issue in the top-level conceptualization of the system. Suppose that for various reasons the communication method of choice is a laser cross-link. Because of the uncertainties involved in this technology, it may be necessary to perform a significant level of detailed analyses at relatively low levels in the system hierarchy in order to mitigate perceived risks with such a concept. How accurately must the laser be pointed? Will it be possible to point the laser to that degree of accuracy? How much onboard electrical power will be necessary to support the cross-link? It may not be possible to answer these questions without significant design and analysis during the early stages of development. The extent of design and analysis necessary will be a function of how much risk the program is willing to tolerate. The more risk adverse, the more design and analysis at lower levels will be required to mitigate perceived risk.

C. Time-Phased Outputs

Another feature of the Time Domain view is that it defines which particular outputs are required, at what fidelity, and when on the program timeline. *To reiterate, the basic activities of the SDF do not change over time. However, the outputs of those activities change dramatically over the course of the development.* The output of a structural analysis during conceptual studies will be quite

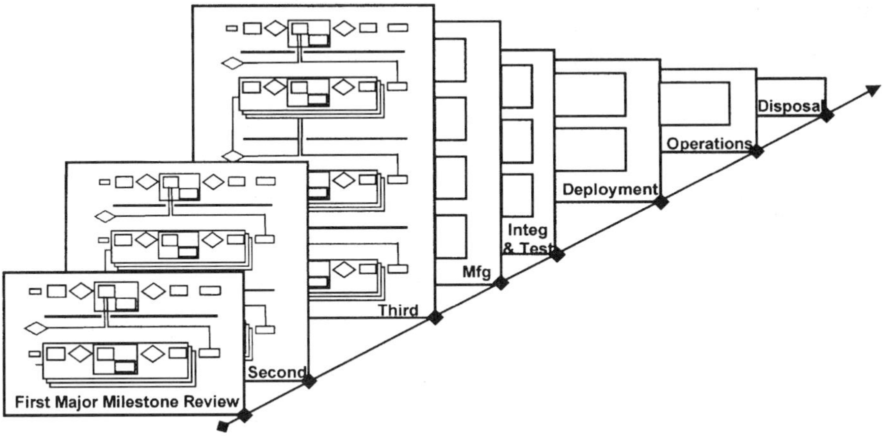

Figure 3.4 Full System Life Cycle.

different from those performed prior to a detail design review. Figure 3.3 indicates that there are two major types of outputs — requirements and design. The figure also illustrates that requirements should "lead" design because design activities respond to requirements. Over time, outputs become more detailed and are generated at increasingly lower levels of the hierarchy.

V. System Life Cycle

Figure 3.4 illustrates the full life cycle for a typical system, addressing how the teams responsible for particular system elements function over time. Supporting teams generally stay intact through Fabrication, Assembly, and Test phases of the program. As the program moves from development to production the composition of the teams may vary widely, moving from an engineering emphasis to production. As the program matures through production, deployment and operations lower-level teams may be subsumed under higher as appropriate.[30]

It is important to consider the full life cycle of the system at the earliest stages of the development because each mission phase imposes unique requirements on the system. In order to maximize the probability of success, these requirements must be considered from the start.

Figure 3.4 further illustrates that the design effort continues up to the final Major Milestone Review that concludes the design phase of the program. After that, the program moves from significant design effort (depicted in the first three "snapshots" with the technical activities shown as per Figure 3.2) to the subsequent phases of manufacturing, integration and test,

[30] Adamsen, Paul B., Jr., Controlling The Chaos: An Integrated Approach To Managing A System Development Program, "Systems Engineering Practices and Tools," *Proceedings Sixth Annual Symposium INCOSE*, Vol. 1, July 7-11, 1996, Boston, MA, pp. 1093-1100.

deployment, operations, and disposal. For a production program, it is important to provide feedback to the design activity, capturing lessons learned from the deployed systems.

chapter 4

The Rework Cycle

I haven't failed, I've found 10,000 ways that don't work.
—Thomas Alva Edison[31]

The only man who never makes a mistake is the man who never does anything.
—Theodore Roosevelt

I. What Is The Rework Cycle?

The Rework Cycle, as defined in the discipline of System Dynamics[32], is a key concept that must be considered in developing an accurate representation of the processes involved in engineering and managing complex system development programs. It was first developed explicitly by Cooper (1980)[33] and has subsequently been adapted by many other authors.[34] Explaining the Rework Cycle, Richardson and Pugh note:

> Not all work done in the course of a large project is flawless. Some fraction of it is less than satisfactory and must be redone. Unsatisfactory work is not discovered right away, however. For some time it passes as real progress, until the need for reworking the tasks involved shows up. Hence, we are led to suggest that the project workforce accumulates two kinds of progress: real and illusory. We shall term the latter undiscovered rework. Together, cumulative real progress and undiscovered rework constitute what is

[31] The author also found some sources that attribute this to Benjamin Franklin.

[32] System dynamics was pioneered by Jay Forrester. His first book on the subject is *Industrial Dynamics*, Portland, OR: Productivity Press, 1961.

[33] Cooper, Kenneth G. Naval Ship Production: A Claim Settled and a Framework Built, *Interfaces*, 10:6, December 1980.

[34] Ford and Sterman note several papers: Cooper, 1993a, b, c, 1994; Seville and Kim, 1993; Jessen, 1990; Kim, 1988; Pugh and Richardson, 1981; Abdel-Hamid, 1984; Ford and Sterman, 1997; Ford, 1995.

> perceived within the project as actual cumulative progress. . . .[35]

Understanding the Rework Cycle is key to understanding the dynamics of complex system development. In the context of simulating the dynamics of a real software development program, Cooper comments:

> . . . we had to simulate the performance of actual projects as they really did occur — not just how they were planned to go, or how they should have gone. . . . To do so we had to explicitly address their substantial rework, as well as its causes, detection, and execution.[36]

The basic rework cycle is depicted in Figure 4.1. It begins with a quantification of the work that must be performed, much like developing a cost estimate based upon an understanding of the scope of tasks involved. All of the planned tasks are in a bucket, as it were, waiting to be performed. The work generation activities begin to expend energy and resources to perform the tasks in the "Work to Do" bucket. Once the work is performed it is allocated to the "Work Done" bucket. But herein lies the problem: some of the tasks that are thought to be done, are not actually done. Some fraction of the work thought to be done must be redone. That portion of the work goes, by definition, into the undiscovered rework bucket.[37]

The quality of the work performed is a key factor that drives the amount of rework in any given System Development activity. Quality, in this context, is a measure of that portion of work performed that is actually completed, as opposed to that which is thought to have been completed. Because the quality of the work is not perfect, some of the work that is thought to be done is not actually done. That work is allocated to the rework bucket.

There is a further difficulty as well. Exactly which tasks are done and which need to be redone? In other words, some of the rework is known: that is, some tasks are known to be incomplete or otherwise problematic. However, as Richardson and Pugh note in the above quote, there will likely be tasks that certainly require rework but which have not yet been discovered. The rework bucket contains both known and undiscovered rework.

Known rework can be intelligently managed because it is known. However, undiscovered rework injects risk into the program precisely because it is not known. It is not a trivial task to manage something that is not known. What is the potential cost, schedule, and technical impact? How extensive

[35] George P. Richardson and Alexander L. Pugh, *Introduction to System Dynamics Modeling*, Portland, OR: Productivity Press, 1981, pp. 56-57.

[36] Cooper, Kenneth G. and Thomas W. Mullen, Swords and Plowshares: The Rework Cycles of Defense and Commercial Software Development Projects, *American Programmer*, May 1993, p. 42.

[37] The buckets labeled "Undisc RW" in Figures 4.1 and 4.2 contain both known and undiscovered rework. There is no need to model these separately since the tasks in each bucket must be redone and are thus handled in the same way in the model.

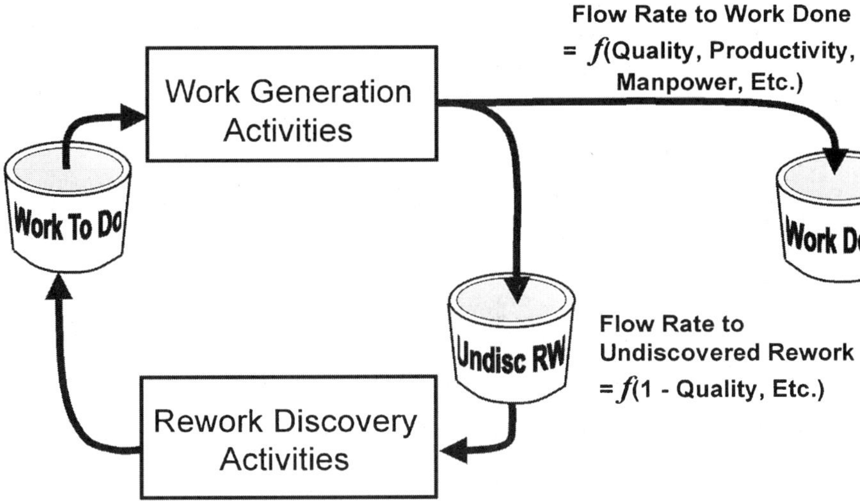

Figure 4.1 The Rework Cycle.

is it? What is the risk to the company and to the customer? Many more issues could be raised, but the key point here is that undiscovered rework is a potentially dangerous thing for all stakeholders. The implication to the system development process is this:

Key Point

> Specific activities must be included in the SDF that are aimed at exposing undiscovered rework in order to mitigate its potentially adverse effects.

Figure 4.1 illustrates a single rework cycle or "phase." The term "phase" in this context refers to a complete rework cycle, unlike its previous usage where it describes different brackets of time along a program timeline. Figure 4.2 depicts a system made up of multiple Rework Cycles, or multiple phases. For simplicity, the activities that compose the Requirements Development activity are coalesced into one phase; similarly, those composing the synthesis activity. Thus, as illustrated here, the system-level portion of the development comprises two Rework Cycle phases. The subsystem tier, which comprises similar Requirements Development and Synthesis activities is depicted with a single Rework Cycle phase. This illustration shows that the fidelity of the output of Requirements Development is a function of the quality of the work performed in its set of activities. This output becomes the input to the system-level Synthesis activity. The quality of the output of the Synthesis activity is limited by the quality of the input data.

The objective here is to illustrate how poor quality ripples through the entire system development in a non-linear fashion. For the purposes of this

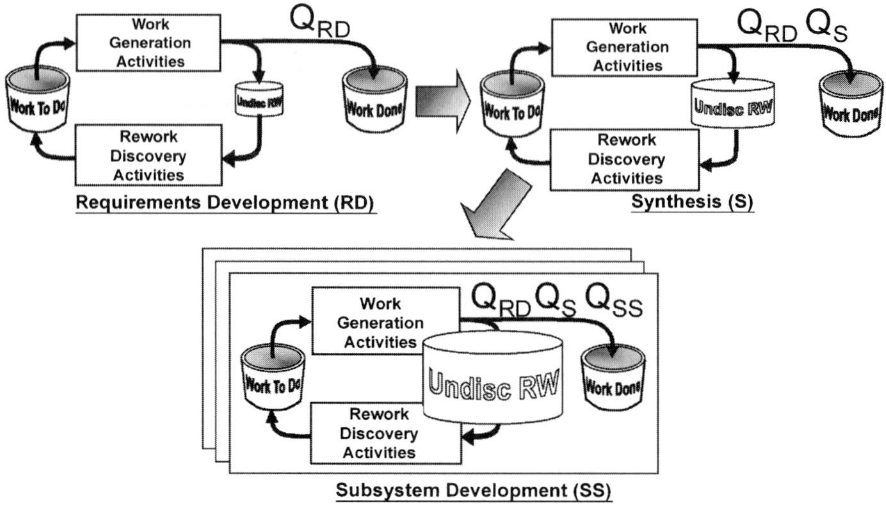

Figure 4.2 The Rework Cycle in Multiple Phases.

illustration, only three phases are shown for an entire System Development program. The rationale for the three phases is that requirements drive the Synthesis activity and the Synthesis activity drives the lower-level development activities. In reality, each sub-activity could be modeled as a rework cycle, each with its own quality level. In a real program there would likely be many more levels in the hierarchy.

It is apparent that even small lapses in quality at upper levels in the hierarchy can have a devastating effect on the quality achievable at lower levels in the hierarchy. Because of feedback effects, this influences quality at upper levels as well. A vicious cycle begins to emerge.

Key Point

> It is clear from this discussion that improving the quality of work performed should be a major priority in sound program management. The amount of work performed that is actually complete is directly related to the quality level of the work. As the quality level goes down, the amount of rework that will be required necessarily increases at a non-linear rate.

Key Point

> This discussion highlights a key difference between complex system development and relatively simple product development. In complex systems, rework at lower levels grows exponentially. The impact of this effect is not as significant in relatively simple products.

II. A Simple System Dynamics Model

In order to illustrate the impact of various parameters on the success of a system development program, a simplified System Dynamics model will be employed. The model includes three phases as mentioned in the discussion concerning Figure 4.2. Some of the key parameters include: quality of the work performed, productivity of the workers performing the task, and the level of effort applied to discovering rework as a percentage of the total staff level. Certainly, more detail could be added with profit, but for the purpose of this discussion, this model will be adequate. When the output is examined, a focus on the absolute numbers is not the primary concern; rather, the central issue has to do with the trends they indicate.

In this context, productivity is simply the rate at which the work is being accomplished. It is a measure of how efficiently allotted time is being used. Quality, in this context, refers to the amount of work actually completed as a fraction of the amount of work thought to be complete. As discussed above, not all the work performed is actually complete. Some fraction becomes either discovered or undiscovered rework.

In the System Dynamics model three variables for quality are employed. The first is "nominal quality," $Q_{nominal}$, which represents the quality of the work as it is being performed by the people or machines. In the model, this number does not change. It is assumed that a fixed percentage of the work performed goes to the rework bucket. The second variable for quality is called "average quality." Once the program starts, it is calculated. Average quality, Q_{avg}, is defined as:

$$Q_{avg} = \frac{Work\ Done}{Work\ Done + Rework} \tag{4.1}$$

As the program progresses, Q_{avg} approaches unity because the work done bucket is filling up and rework is being discovered and thereby eliminated from the rework bucket. The third variable is quality, Q. It is a function of the nominal quality of the phase in question times the average quality of the input data. Thus, each phase starts at a quality level, Q, equaling its nominal quality times the quality of the input data. This is illustrated in Figure 4.3 which is an output of the System Dynamics model. The horizontal axis represents time, while the vertical axis represents the quality level on a scale of zero to one.

The initial quality level, $Q_{nominal}$, for each phase was defined as 50%. The quality of the first phase remains constant because, in this simplified model, it is not dependent upon downstream phases for its input.[38] However, the quality level of the output of the first phase is the average quality level or $Q_{Phase\ 1\ avg}$. The quality level of the Phase 2 activity is calculated as follows:

[38] It is recognized that this is a simplification. In real life there would be feedback from the downstream activities. Such feedback is developed in Chapter 5.

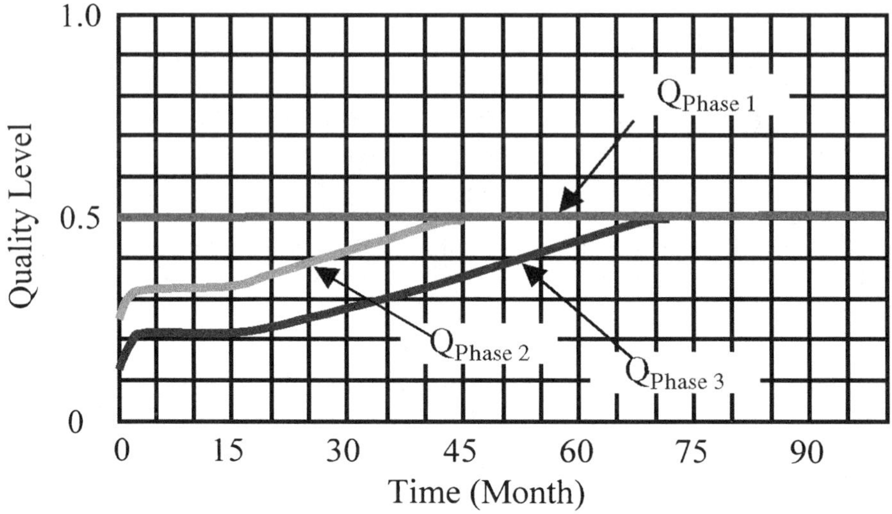

Figure 4.3 Dynamics of Quality over Time.

$$Q_{Phase\ 2} = Q_{nominal} Q_{Phase\ 1\ avg} \tag{4.2}$$

This is because the Phase 2 activity is dependent upon the output of Phase 1 for its input. The instantaneous quality of the output of Phase 1 is its average quality at the particular instant. Therefore, as shown in Figure 4.3, the initial quality level for the Phase 2 activity is 25% (50% × 50%). It rises to the nominal value over time as rework is discovered in Phase 1 and converted to work complete, thus raising the quality level of the input to Phase 2. Likewise, for Phase 3 the initial value is 12.5% (50% × 50% × 50%).

Key Point

> The obvious assumption here is that the Phase 1 activity continues to discover rework in order to convert it to work complete. If this does not happen, the quality level of the input to Phase 2 does not improve and the quality level of the Phase 2 work stagnates at 25%. This represents a classic "Pay me now or pay me later" scenario. Poor quality work that is not corrected early will cause many orders of magnitude more serious problems in the future. In terms of the model, if the average quality of Phase 1 is not improved, the model will not converge — it will not be able to finish because not all the work is complete. How much more on a real program?

Key Point

> This discussion highlights an issue regarding the optimal timing of the release of output data from one activity as input data to subsequent activities. Should the data be input to the next activities as soon as it is available or should the rework activities be given time to improve the data? The answer to this question involves an assessment of the quality of the output data. The better the data, the sooner it makes sense to pass it downstream. The poorer the quality of the data, the more rework it will generate in subsequent activities and thus increase overall program costs (Garbage in → Garbage out).

Returning to Figure 4.3, it can be seen that as more phases are added, or the more tiers in a complex hierarchy, the more rapidly quality deteriorates at lower levels. It can also be seen that the schedule is pushed farther to the right. This is apparent since the horizontal axis represents time and the phase is not completed until the quality level reaches its nominal value. The quality level of Phase 3 at any given instant in time is calculated as follows:

$$Q_{Phase\ 3} = Q_{nominal} Q_{Phase\ 1\ avg} Q_{Phase\ 2\ avg} \tag{4.3}$$

where the values of $Q_{nominal}$, $Q_{Phase\ 1}$, and $Q_{Phase\ 2}$ are the instantaneous values at the time $Q_{Phase\ 3}$ is calculated. If k equals the number of phases in the hierarchy, then a pattern emerges whereby:

$$Q_{Phase\ n+1} = Q_{nominal} \left(\prod_{n=0}^{k} Q_{Phase\ n\ avg} \right) \qquad (4.4)^{39}$$

If all the phases have the same quality level, as it is assumed in this model, then:

$$Q_{Phase\ n+1} = Q_{nominal} Q_{avg}^{\ n} \tag{4.5}$$

Of course, Equation 4.5 is only true for the initial value of quality since after time equals zero, each Q_{avg} changes at a different rate.

Exponential Rework Growth — The aforementioned System Dynamics Model is used here to briefly examine the effects of rework as it ripples

[39] Equation 4.4 assumes that for $n = 0$, $Q_{Phase\ n\ avg} = 1$, such that $Q_{Phase\ n+1} = Q_{Phase\ 1} = Q_{nominal}$

through this simple three-phase hierarchy.[40] Figure 4.4 depicts the amount of rework generated by each phase as a function of time, with a nominal quality value of 50%. For Figures 4.4 through 4.6 the horizontal axis represents time, and the vertical axis represents the number of tasks waiting to be performed. It is clear from the figure that rework is growing in a non-linear fashion. It is also clear from this view that the schedule is being pushed to the right in a significant way. This is indicated because each phase is completed when all of the tasks have been completed; in order for all the tasks to be completed, all of the rework must be emptied from the rework bucket. Thus when the rework bucket remains empty, the phase is completed.

Note that as the results of the simulations performed are presented, both cost and schedule impacts are identified. The schedule impacts are observed directly from the figures since time is the unit of measure along the horizontal axis. Cost is a function of the number of tasks that must be performed. Therefore, an increase in the number of tasks that must be performed indicates a commensurate increase in cost. Such is the case with rework. Rework represents tasks that were performed at least once, but for various reasons, must be redone.

Note in the following discussion that the effort applied to generating work, or the Potential Work Rate (units = tasks per month), is fixed by the number of staff (units = persons) available and their respective productivity (units = tasks per month per person). For all simulations the number of staff generating work is fixed at five people. The potential work rate for Rework Discovery activities is determined by multiplying the Potential Work Rate for work generation activities by the fixed percentage identified in each simulation. For example, if the effort applied to Rework Discovery activities is set at 20% of the effort applied to generating work, then the effort applied to discovering rework would be 20% times five people times the productivity level.

Figure 4.5 provides a similar view of the program, but with a nominal quality level of 70%. The contrast between these two emphasizes the need to minimize rework generated. Rework still cascades non-linearly through the hierarchy. However, its magnitude is reduced significantly as quality is improved.

Figure 4.6 illustrates rework growth at a quality level of 90%. Notice that the scale of the vertical axis has been changed relative to Figures 4.4 and 4.5. Both Figures 4.4 and 4.5 have a vertical axis scale of zero to 40. Figure 4.6 has a vertical axis scale of zero to 4, indicating that the rework generated with a quality level of 90% is negligible as compared to quality levels of 70% and 50%. Notice also the schedules of the three figures. Completion of the phase is indicated when the bucket containing the remaining undiscovered rework tasks reaches zero. With a quality level of 90% the program finishes

[40] See Appendix F for a description of the model itself. The effects of schedule pressure, staffing, etc. have not been modeled.

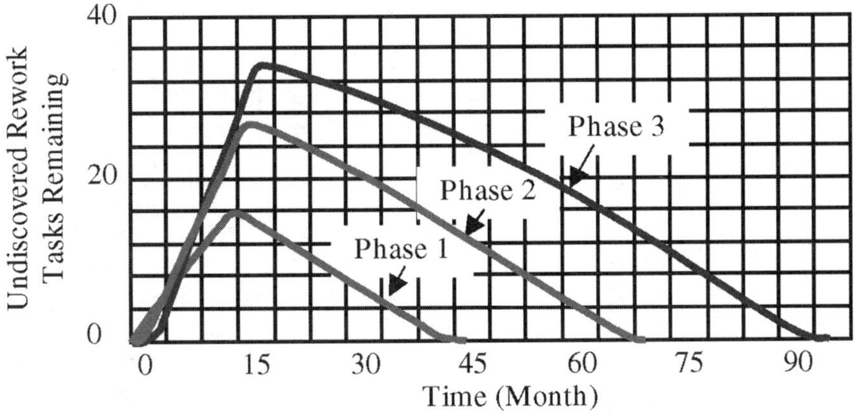

Figure 4.4 Rework Growth as a Function of Number of Phases (Quality = 50%).

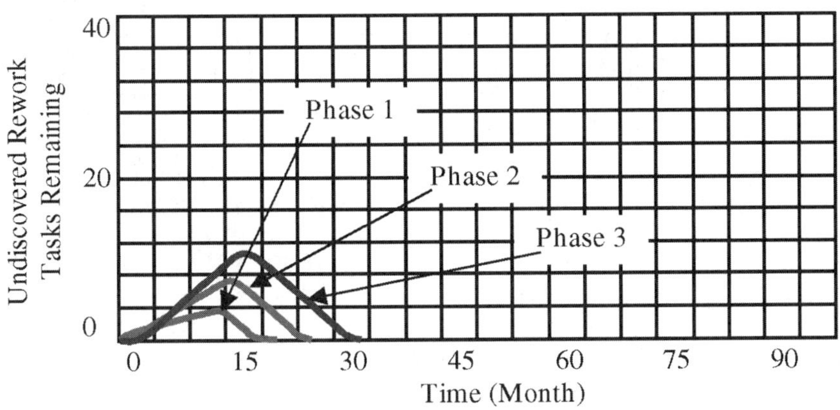

Figure 4.5 Rework Growth as a Function of Number of Phases (Quality = 70%).

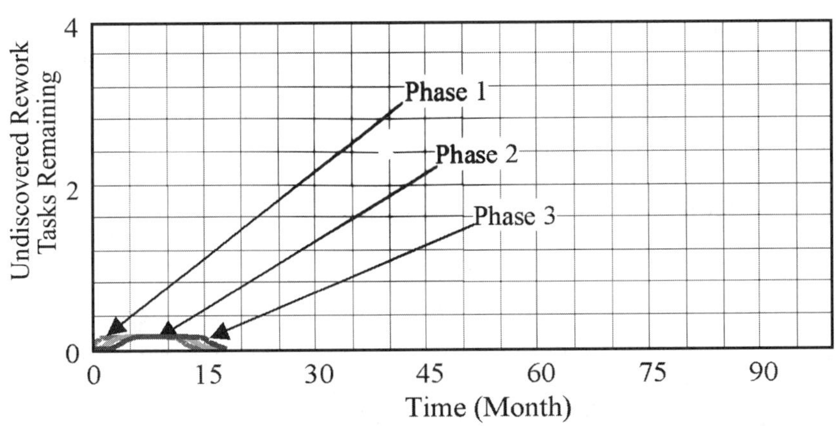

Figure 4.6 Rework Growth as a Function of Number of Phases (Quality = 90%).

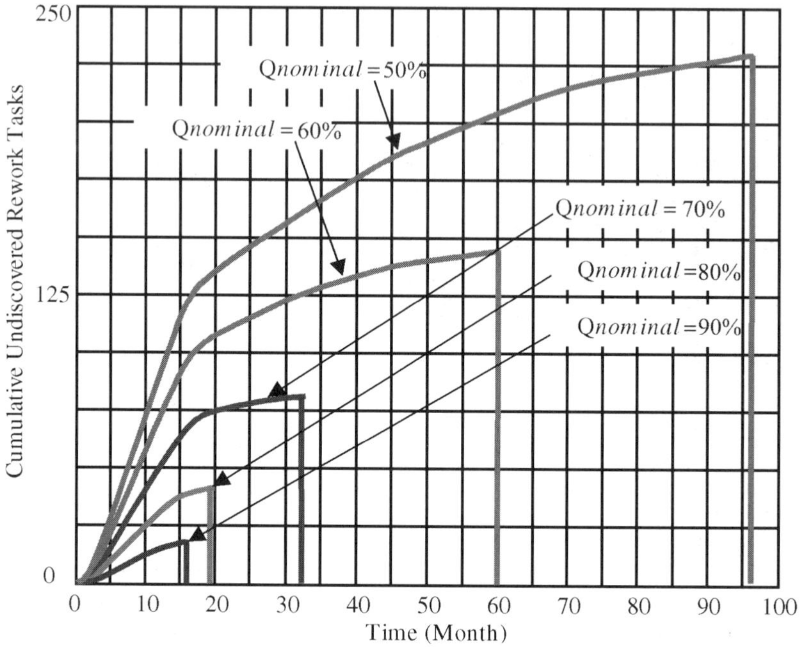

Figure 4.7 Cumulative Rework Generated as a Function of Quality Level.

in about 17 months. With a quality level of 70% the program finishes in about 32 months. At a quality level of 50% the program finishes in about 96 months.

The absolute numbers here are not important. This is a relatively simple model and a real program is far more complex. Nevertheless, the trends the numbers indicate are worthy of note. It is obvious from this analysis that the number of tasks that need to be reworked is significantly reduced by increasing the level of the quality of work performed. It naturally follows that, with less work needing to be redone, the program can be completed much faster and therefore in a much more cost-effective manner.

Figure 4.7 provides a comparison of how the cumulative amount of rework generated grows non-linearly as the quality level deteriorates. The vertical axis represents the cumulative number of Rework Discovery tasks generated. The same point made in Figures 4.4, 4.5, and 4.6 above is reiterated here: Rework resulting from poor quality adds significant cost to the program in terms of both dollars and schedule slippage.

Key Point

> Rework can also induce technical risk to the program because the results of the corrected work could involve changing technical parameters that invalidate certain aspects of the design.

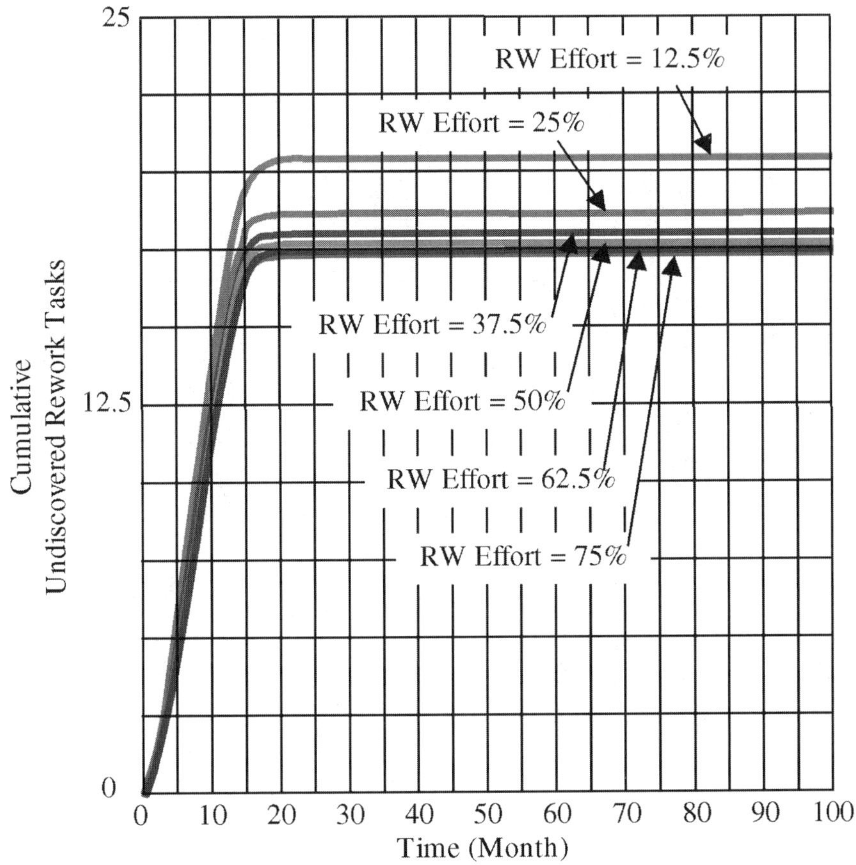

Figure 4.8 Rework Generated as a Function of Rework Discovery Effort — Quality = 90%.

The SDF outlined in this book identifies specific tasks that ought to be performed specifically for the purpose of discovering rework as early in the process as possible. This is necessary in order to mitigate its potential negative consequences.

How Intense Should Rework Discovery Be? — As mentioned above, in this simulation, the amount of effort applied to discovering rework is defined as a percentage of the effort applied in generating work complete. Figures 4.8, 4.9, and 4.10 illustrate the impact of varying the level of effort applied to discovery of rework. For each simulation the productivity level is set to 90%. The horizontal axis in each figure represents time, while the vertical represents the total number of rework discovery tasks generated (more is worse).

In Figures 4.8 through 4.10 it is the knee in the curve that indicates at what point Rework Discovery activities level off. This is the point at which the phase concluded. In order to find the optimal percentage of effort that

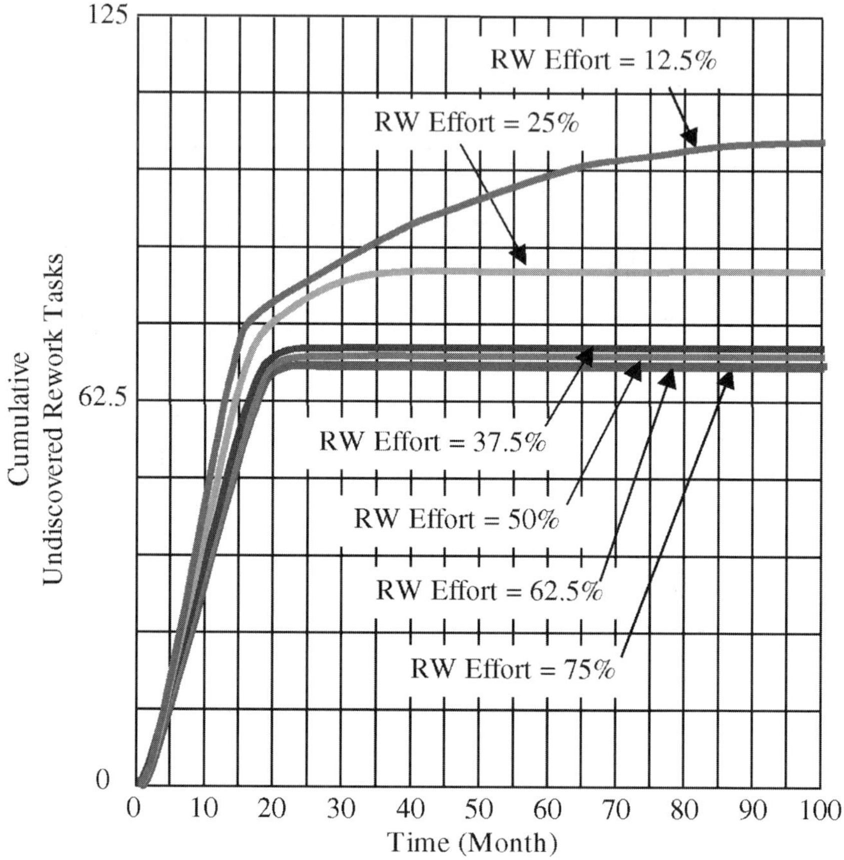

Figure 4.9 Rework Generated as a Function of Rework Discovery Effort — Quality = 70%.

should be directed toward discovering rework, the curve with the lowest cumulative amount of rework must be identified. For the first simulation, the nominal quality parameter is set to 90%. At this quality level, there is minimal difference between applying 25% effort toward discovering rework and applying higher percentages. However, as quality deteriorates, more effort must be applied to discover rework in order to minimize its impact to cost and schedule.

Figure 4.9 shows the impact of operating at a quality level of 70%. At such a quality level, about 38% of the effort applied should be directed toward rework discovery. More than that would not be cost effective. Less than 38% would result in unnecessarily high cost and schedule impacts.

Figure 4.10 shows the results of running the simulation at a quality level of 50%. In this scenario, the effort applied to discovery of rework should be on the order of 50% or more.

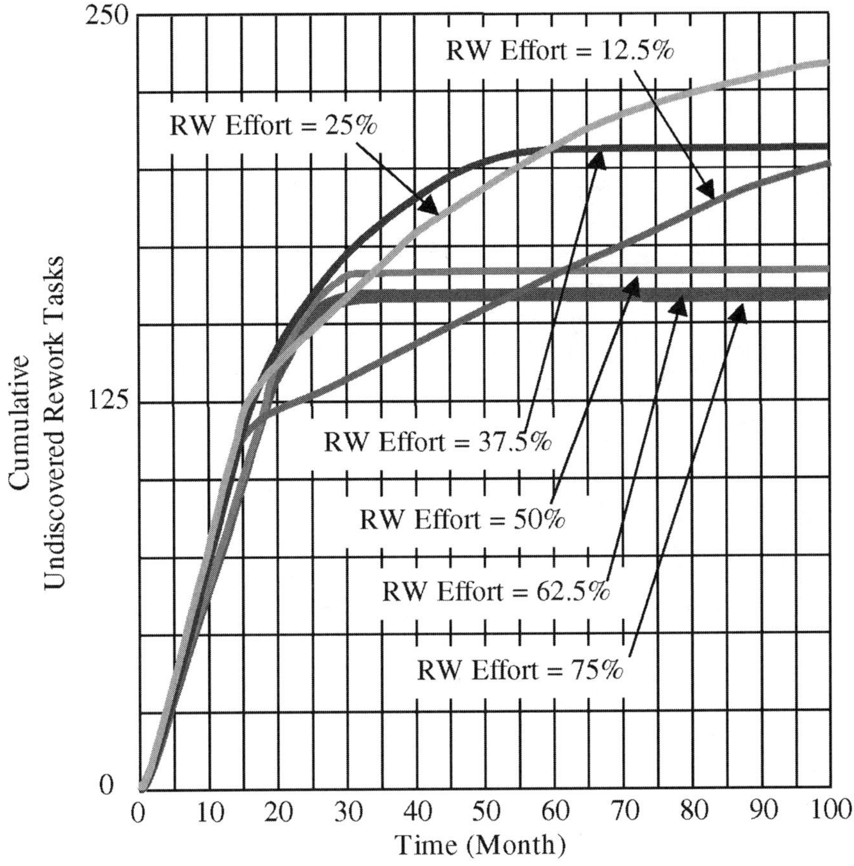

Figure 4.10 Rework Generated as a Function of Rework Discovery Effort — Quality = 50%.

Looking at Figures 4.8 through 4.10, notice how the total rework generated grows at a non-linear rate as quality deteriorates from 90 to 70% and finally to 50%. At a quality level of 90% (Figure 4.8) the total rework ranges from about 17 to 21 tasks, depending upon how much effort is applied to finding it. At a quality level of 70% (Figure 4.9) the rework grows to a minimum of about 60 tasks. At a quality level of 50% (Figure 4.10) the rework grows to a minimum of about 160 tasks. In terms of schedule, at 90% the program finishes within about 20 months; at 70% it finishes in about 25 months; and at 50% it completes in about 35 months when the rework discovery effort is run at the optimum level. Notwithstanding the above, the main point of Figures 4.8 through 4.10 is not to determine the optimal amount of rework discovery effort that must be applied in those specific situations.

Key Points

- Failure to acknowledge the existence of rework and to address it properly can be detrimental to a development program.
- Poor quality results in unnecessarily high cost and schedule resources spent on fixing problems that may well have been avoided.
- The lower the quality level, the higher the percentage of resources must be applied to discover the resulting rework in order to minimize its adverse effects on cost and schedule. This relationship between quality level and resulting rework is exponential.

III. Rework Mitigation

The preceding discussion has shown how even small degradations in the quality of work performed can undermine the ability of a program to operate within budgeted cost and schedule constraints. Undiscovered rework injects cost, schedule, and technical risk into any development program because the nature of the problems is not known. Undiscovered rework, by definition, is unwittingly accepted within the program as work complete. Such data is used to make design decisions, make-buy decisions, critical trade-offs, cost estimates, schedule estimates, etc. This undiscovered rework will make itself known at some point in the program: during subsequent design phases, during manufacturing, during integration and test, during deployment, or during operations. It is obvious that the later in the program it is discovered, the more difficult and expensive it will be to rectify. In addition, there may be hazards to humans and the environment hidden in the design.

Key Point

> The main point here is that improving the quality of work performed must be a priority if a program is to be run at minimal risk. As a corollary, effort must be made to discover undiscovered rework in order to minimize its cost, schedule, and technical effects on the program.

What are some of the causes of rework during the development process? The following nonexhaustive list identifies some of the sources.

- Imposed requirements incomplete, ambiguous, contradictory, unverifiable
- Undiscovered mission requirements

- Planned technology not available in time
- Initial design "granularity" insufficient to detect fundamental issues
- Heritage hardware and/or software not robust enough for new context
- Interfaces not compatible
- Initial technical, cost, and schedule allocations not sufficient
- Inadequate control and flow of requirements and information
- Incomplete or too top-level performance analyses
- Testability and/or producibility difficult and expensive
- Failure to adequately test and/or simulate design
- Inadequate test planning

These are not uncommon causes of rework. This short, nonexhaustive list suggests that activities ought to be included in the SDF that are specifically aimed at discovering these causes of rework. In Adamsen (1995), a linear, sequential SDF is developed. As Table 4.1 indicates, the activities identified, which are typical of most system engineering processes, fall naturally into two categories: those activities focused on generating work and those focused on discovering rework. It is interesting that one half of those activities focus on the discovery of rework.

Key Point

> Sometimes the activities described in the table as focused on rework discovery are viewed as "non-value-added" activities. The preceding discussion shows that such a view is misguided. There is much "value-added" to finding rework early, which will serve to ensure that the program proceeds with minimal technical, cost, schedule, and safety risk to the program.

Table 4.1 Focus of SDF Activities Defined in Adamsen (1995)

Requirements Development		Design and Analysis				Verification	
Activity	Main Focus	Activity	Main Focus	Activity	Main Focus	Activity	Main Focus
Requirements Analysis	Discover Rework	Identify/Modify Design	Work	Analyze Performance	Discover Rework	Analysis (may be same as Synthesis) Test	Discover Rework
Mission Analysis	Work	Allocation	Work	Assess Producibility, Testability	Discover Rework		Discover Rework
Requirements Verification Check	Discover Rework	Functional Decomposition	Work	Optimize	Work	Plan System Test	Discover Rework
Functional Analysis	Work and Discover Rework	Design Integration	Work				

chapter 5

System Development Framework — Technical

It is not the critic who counts, not the man who points out how the strong man stumbled, or where the doer of deeds could have done them better. The credit belongs to the man who is actually in the arena; whose face is marred by dust and sweat and blood; who strives valiantly; who errs and comes short again and again; who knows the great enthusiasms, the great devotions, and spends himself in a worthy cause; who, at the best, knows in the end the triumph of high achievement; and who, at worst, if he fails, at least fails while daring greatly, so that his place shall never be with those cold and timid souls who know neither victory nor defeat.
—Theodore Roosevelt

In this chapter, a detailed decomposition of the System Development Framework (SDF) building block is developed, which was discussed in the preceding chapters. An example, loosely drawn from a conceptual spacecraft study done in the early 1990s, is included in order to clarify the application of the SDF. The primary focus of the example is functional analysis and decomposition. The spacecraft mission from which this example was drawn was the operation of an astronomical telescope in a low earth orbit. The example is simplified in order to keep attention focused on explaining the application of the SDF to a real development activity. In order to maintain clarity of the SDF discussion, the example will be applied at the end of each section. This spacecraft system will be referred to as Example Sat or ESAT for short. The task for the ESAT program is to develop the spacecraft bus — that portion of the satellite that supports the telescope in space.

As discussed in Chapter 1, the term "system" has been defined as "any entity within prescribed boundaries that performs work on an input in order to generate an output." Using this definition, it is asserted that at any level of the development hierarchy there exist one or more "systems." At the level below the top level, or system level, these are commonly called subsystems.

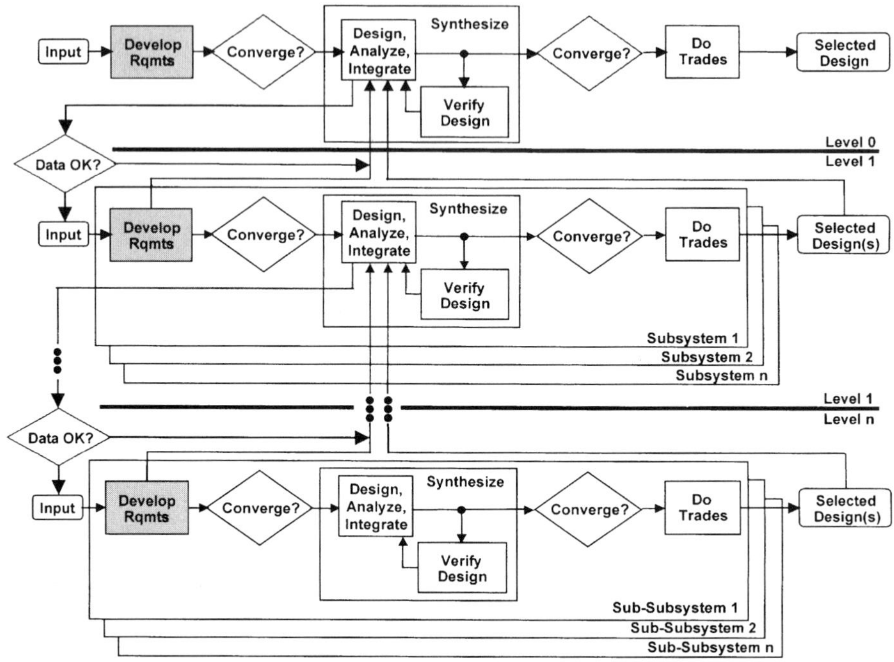

Figure 5.1 "Develop Requirements" in the System Hierarchy.

Each of these subsystem development activities is perpetuated as its own "system," applying the structured approach herein described as the SDF. Therefore, it is further asserted that while the specific inputs and outputs may change with hierarchical levels, the general activities delineated in this chapter do not change at each level. Thus this process is applicable to all levels of the hierarchy. In the discussion below, it is assumed that each of the activities described is applied to each development activity at each respective hierarchical level.

In order to keep the following discussion in the context of the total SDF, Figures 2.2 and 3.2 from the preceding chapters will be used in various sections to highlight the activity under discussion.

I. Develop Requirements — Determine "What" the System Must Do

Figure 5.1 highlights the "Develop Requirements" activity at each level in the system hierarchy. This serves to illustrate that the same basic process is applied to each system element at each hierarchical level.

The Requirements Development set of activities addresses the question, "***What*** must the system do?" These activities include:

- Collect and analyze imposed requirements
- Derive requirements through context analysis, functional analysis, design, allocation, and decomposition
- Manage requirements derived during the development process
- Communicate requirements and requirements changes
- Determine and track how and where in the system build-up the requirements will be verified
- Achieve customer consensus regarding interpretation of the customer-imposed requirements
- Maintain traceability of requirements

It is important to maintain traceability of requirements throughout the development for several reasons:

- Cost Minimization — Avoid over-design; that is, adding cost by including functionality that is not necessary; or under-design, i.e., not providing functionality that is required by the customer
- Cycle Time Reduction — Facilitates a coordinated effort that "does it right the first time"
- Change Impact Analyses — Provides a logical and systematic approach to assessing the impacts of changes to the design
- Customer Requirement — Many customers require demonstrated traceability
- Consistency, Clarity, and Completeness Ensured — Early detection and correction of requirements issues

Figure 5.2 illustrates the decomposition of the Requirements Development (RD) activity, derived in Chapter 2, Figure 2.2. The RD activity is organized as a rework cycle as discussed in Chapter 4. It comprises two work generation activities: "Derive Context Requirements" and "Generate Functional Description," as well as two rework discovery activities: "Analyze Requirements" and "Analyze Functional Description."

As the RD activity progresses, it is continually or periodically assessed for convergence. Convergence occurs when the design team is able to reach a reasonable solution with the input data. There are several indicators that can be employed to assess if convergence is occurring: monitoring key Technical Performance Measures (e.g., technical budgets, remaining margin on key parameters), program risk assessments, various technical analyses, estimates to complete, and periodic design reviews and audits. If convergence is not occurring at an acceptable rate, changes may be required to enhance the quality of the work performed, and/or to accelerate the rework discovery activities.

RD rework discovered elsewhere in the process may feed back to the RD activity as "work to do" or as issues to be addressed with the customer who generated the input requirements. This is indicated by the "discovered

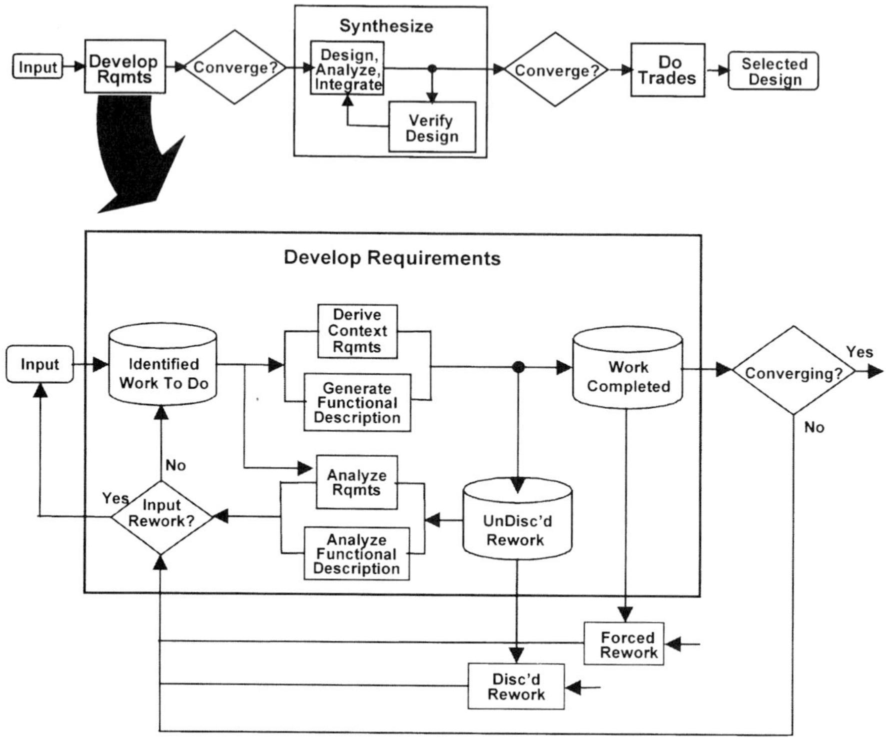

Figure 5.2 The "Develop Requirements" Activity Decomposed.

rework" box. Rework discovered in other areas may be the result of difficulty implementing the requirements as defined in the functional description. This may force work previously defined as complete to be redone. This is indicated by the "forced rework" box.

A. Inputs

Requirements originate from many sources in varying forms, both explicit and implicit. These include technical, cost, and schedule concerns. All requirements must be considered to maximize success. The following is a nonexhaustive list of potential requirement sources that ought to be considered.

- Immediate customer
- The division
- Business development
- Subcontractors
- Procuring organization
- The corporation

- Heritage designs
- New technology
- User community
- The department
- Competitors

Some of the properties that compose a good requirement include:

- Clarity — unambiguous
- Consistency — no mutually exclusive or conflicting requirements
- Completeness — provides all necessary information
- Verifiability/Quantifiability — compliance demonstrable
- Traceability — necessity demonstrable
- Functionally oriented — maximizes design creativity/flexibility

The requirements management discipline has a nomenclature of its own. Some of the key terms are:

- Parent — A requirement from which other requirements have been derived
- Child — A requirement derived from a parent
- Orphan — A non-top-level requirement having no identified parent (otherwise described as a problem child)

Each of these are illustrated in Figure 5.3.

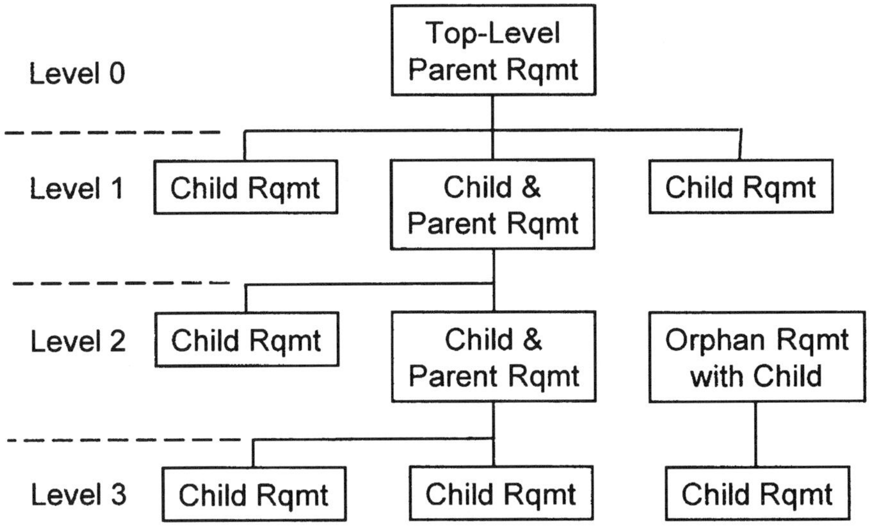

Figure 5.3 Requirements Relationships.

Other important terms in a requirements document include: shall, will, and may or should. In most contexts, employment of the term "shall" indicates that there is no flexibility in terms of the design providing that particular function and that function performing according to the specified level. Where there is some flexibility regarding the customer's expectations, other words are generally used such as "will," "may," or "should." Therefore, it is important for all the stakeholders to define these terms up front and to use them consistently so that any trade-offs can be performed according to the right priorities.

As discussed above, requirements originate from many different sources. Table 5.1 provides an abbreviated listing of some of the initial requirements defined by the customer at the beginning stages of the ESAT conceptual study. Some of these were recorded in presentation packages provided by the customer, others were mentioned in telephone or other informal conversations.[41] This is not unusual and it is important, even at this early stage in the program, to maintain traceability. Therefore, as shown below, the source of the requirement is recorded with each requirement.

B. Work Generation Activities

Figure 5.4 highlights the Work Generation activities that are performed within the Develop Requirements activity.

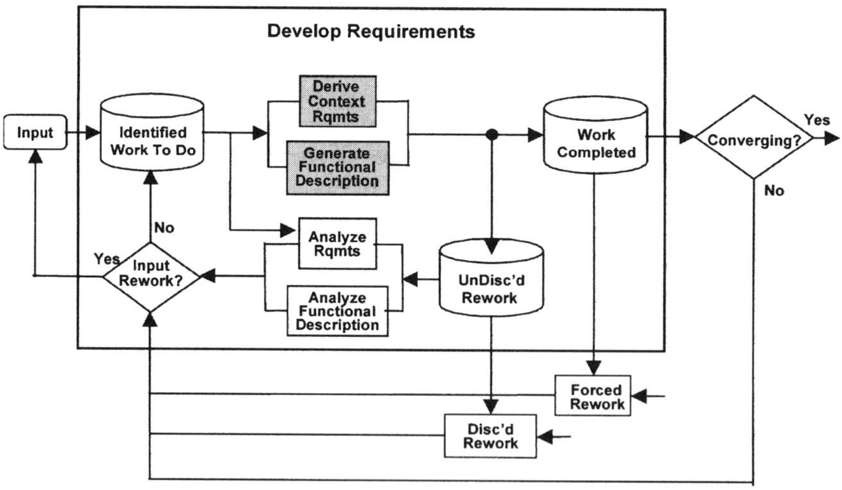

Figure 5.4 "Develop Requirements" Work Generation Activities.

1. Derive Context Requirements

The focus of this activity is to determine context in which the system must function over its complete life cycle. This is accomplished by:

[41] Most of these requirements are taken from the study; some are fabricated to facilitate the usefulness of this example.

Table 5.1 ESAT Customer-Imposed Requirements Set

No.	Title	Text	Source
1.0	**General Program**		
1.1	Launch Date	The spacecraft shall support a launch date of October 1998	Presentation Package
1.2	Operational Orbit	The operational orbit shall be 800 Km altitude, 28 degrees inclination	Presentation Package
1.3	Mission Life	The spacecraft bus shall provide full functionality for a minimum of 2 full years' operation on orbit after initialization	Presentation Package
2.0	**Space Segment**		
2.1	Payload Instrument		
2.1.1	Fine Star Sensor (FSS) Accuracy	The instrument shall provide FSS data to the spacecraft bus with an accuracy of 1/2 Hz. 0.33 arcsec (1 sigma) and a 20 arcmin field of view	Presentation Package
2.1.2	Instrument Mass	The instrument mass shall not exceed 1500 pounds mass	Presentation Package
2.2	Spacecraft Bus		
2.2.1	Instrument Data Interface	The spacecraft bus shall provide a 4 to 300 kbps data interface	Presentation Package
2.2.2	Slew Rate	The spacecraft shall be able to slew 90 degrees within 45 minutes of initialization	Presentation Package
2.2.3	Contamination	The cleanliness level 500A shall be maintained during integration and test of the system	Presentation Package
2.2.4	Command and Telemetry Interface	The spacecraft shall provide MIL-STD-1553 and MIL-STD-1773 command and telemetry data bus interfaces	Presentation Package
2.2.5	On-Board Data Storage	A minimum of 100 megabytes of data storage shall be provided by the spacecraft	Telecon w/ Program Manager
2.2.6	Mechanical Interface	The spacecraft mechanical interface shall be a 4 point attachment on a 48 inch bolt circle	Presentation Package
2.2.7	Electrical Power	The spacecraft bus shall be capable of providing up to 300 watts of power at 28 +/-7 v End Of Life	Telecon w/ Program Manager

Table 5.1 (continued) ESAT Customer-Imposed Requirements Set

No.	Title	Text	Source
2.2.8	Attitude Control	The spacecraft shall point the telescope to an accuracy of ±0.01o on all three axes	Presentation Package
3.0	**Launch Segment**		
3.1	Spacecraft Bus Volume	The spacecraft bus volume shall not exceed 108 inches in diameter and 36 inches height above the separation plane	Presentation Package
3.2	Total Deliverable Mass	The maximum deliverable mass to the operational orbit shall be 3000 pounds	Presentation Package
3.3	Fairing Volume	The maximum fairing volume shall be 108 inch diameter, TBD inches height	Presentation Package
3.4	Minimum First Mode	The minimum first mode shall be 12 Hz	Presentation Package
4.0	**Ground Segment**		
4.1	Antenna Configuration	The ground system antenna shall be a dichroic design, 5 meters in diameter	Telecon w/ Program Manager

- Identifying all mission phases, modes, and states
- Identifying and characterizing all external interfaces, by mission phase
- Defining the environments to which the system will be subjected, by mission phase
- Identifying critical issues by mission phase (events, technologies, etc.)
- Developing the concept of operations

Output

- Specification(s)
- Operations Concept
- Context Diagram(s)
- Entity Relation Diagram(s)
- Event List(s)
- External ICDs

Figure 5.5 illustrates some of the key parameters that define the context within which the system must function over its life cycle. The various phases of the program are identified as columns ranging from "womb-to-tomb." Key parameters are identified as rows in the matrix. This provides an organized framework to begin deriving context requirements.

Parameters	Mission Phases				
	Integration & Test	Deploy	Initialization	Operations	Disposal
External Interfaces	Test Fixtures	Launcher Ground Sys	Launcher Ground Sys AKM	Ground Sys Relay Sats Other Sats	Ground System
Environment	Clean Room System Test	Air Ride Van Air Transport Launch site Facilities Fairing	Ascent Traj	Operational Orbit	Parking Orbit or Earth Re-Entry
System Modes	Test	Test Launch mode	On-Orbit test Maneuver Appendage Deploy	Nominal Standby Safe Maintenance On-Orbit test	De-Orbit

Figure 5.5 ESAT Mission/Context Definition

Key Point

> It is necessary to consider all phases of the program early in the development process because each phase may impose unique requirements to the system. During manufacturing, assembly, and test activities or integration and test activities, for example, special interfaces may be required. This analysis should also expose incompatibilities among certain interfaces. The earlier in the design process that these things can be addressed, the higher the probability the system will be successful.

2. *Generate Functional Description*

As requirements are developed, the Functional Analysis activity seeks to arrange the functions into a coherent system. This can be done in several ways, one of which is computer simulation. A key goal is to ensure that there are no mutually exclusive or conflicting requirements.

This activity generates the specification of the system. A key concern here is proper protocols, and timing of input and output data. Outputs include:

- Identification of all functional requirements flowing out of imposed and derived requirements
- Development of the specification(s)
- Determination of performance requirements of each function and the relationships (interfaces, interdependencies, etc.) between functions

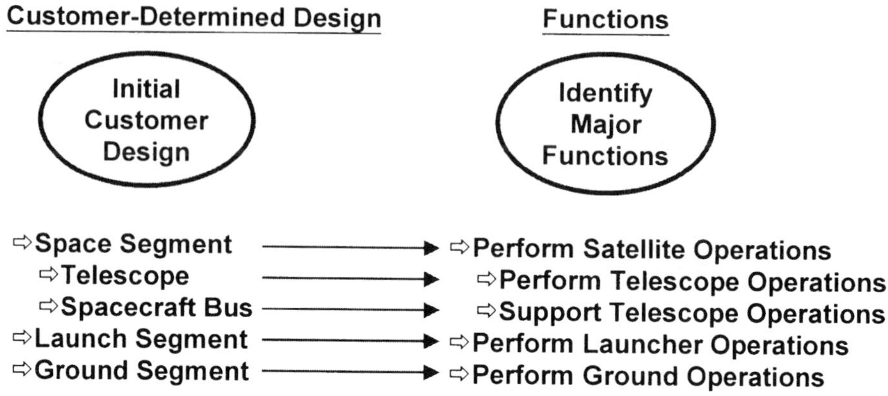

Figure 5.6 Mapping Selected Implementation to Functions.

Key Point

> The way in which the customer defined his system-level requirements reveals his selection of a particular implementation of the system. He has determined that the telescope requires a dedicated spacecraft bus to support it; that a particular launch vehicle will be required; and that a particular ground station configuration will be used.

Figure 5.6 illustrates the mapping of the customer-defined system-level implementation of the program. Each major segment identified in the requirements corresponds to a function that must be performed by the system.

Figure 5.6 illustrates the primary functions that must be performed from a top-level system perspective during the launch and orbit acquisition phase of the mission. Notice that each function description begins with a verb. This is important because it emphasizes the essence of a function — it is something the system must do.

Key Point

> A functional description is not primarily concerned with defining "how" the system ought to be designed. Its purpose is to describe "what" the system must do. Such definition facilitates new ways of implementing systems — thinking "out of the box," which nourishes an environment conducive to design breakthroughs.

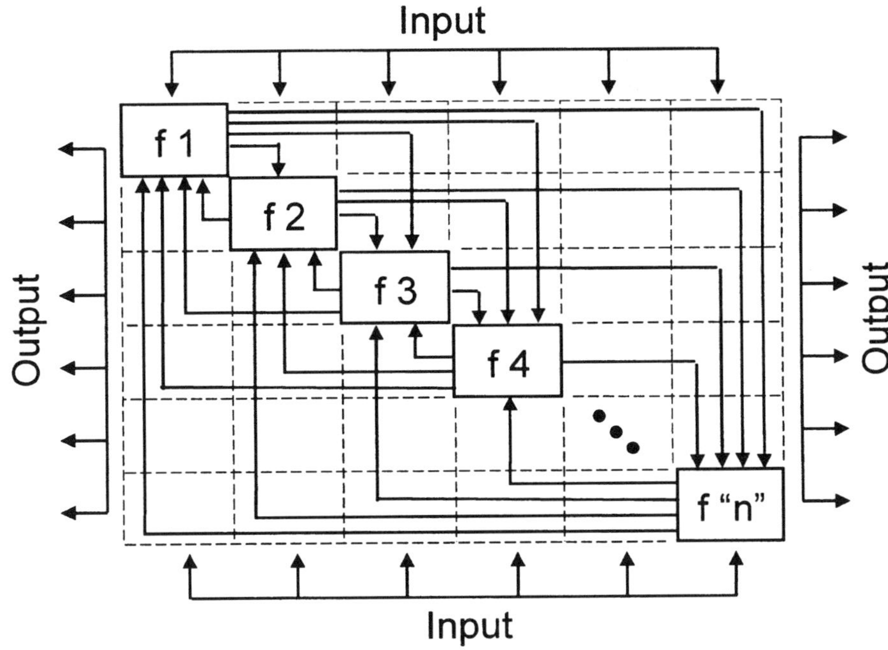

Figure 5.7 The N^2 Diagram.

As the system is developed, the N^2 Diagram format will be loosely followed.[42] Figure 5.7 provides an illustration of the N^2 format, which is a convenient way of developing interfaces between system elements, whether functions or implementation. System elements are placed along the diagonal, inputs and outputs are indicated on the outer-sides of the chart, and interfaces are defined as shown.

During the Orbit Acquisition Phase, the ESAT system comprises three major segments: ground, space, and launch. Figure 5.8 illustrates the derivation of three major system functions from these major pieces of the system: Perform Satellite Operations, Perform Launcher Operations, and Perform Ground Operations. In keeping with the N^2 format, the functions are arranged diagonally with interfaces, inputs, and outputs described as shown.

Figure 5.9 illustrates the importance of defining system functions for each distinct phase. While both Figure 5.8 and Figure 5.9 describe the ESAT program at the same system level, they are quite different. Figure 5.9 does not have the same functions or interfaces as those shown in Figure 5.8. During the Launch Phase, the Launch Operations function is a major function with critical interfaces identified. Obviously, after the spacecraft achieves its operational orbit, the Launch Operations function is no longer required.

[42] Cf. the *Systems Engineering Management Guide*, January 1990, Section 6.3.2 for a discussion of the N^2 Diagram. As noted in that reference, it was developed by TRW and is described in "The N^2 Chart," R. Lano, Copyright 1977 TRW Inc.

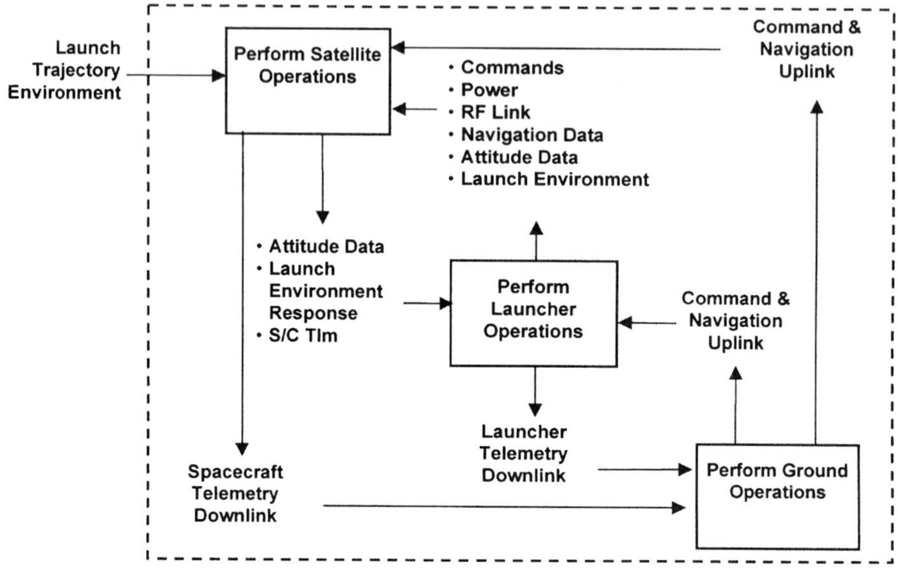

Figure 5.8 ESAT System-level Functional Block Diagram — Orbit Acquisition Phase.

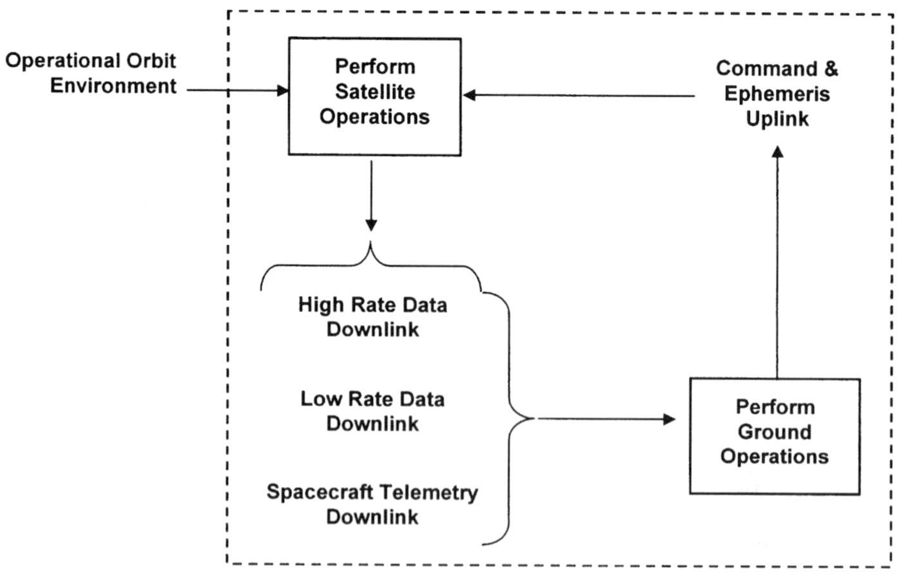

Figure 5.9 Operations Phase.

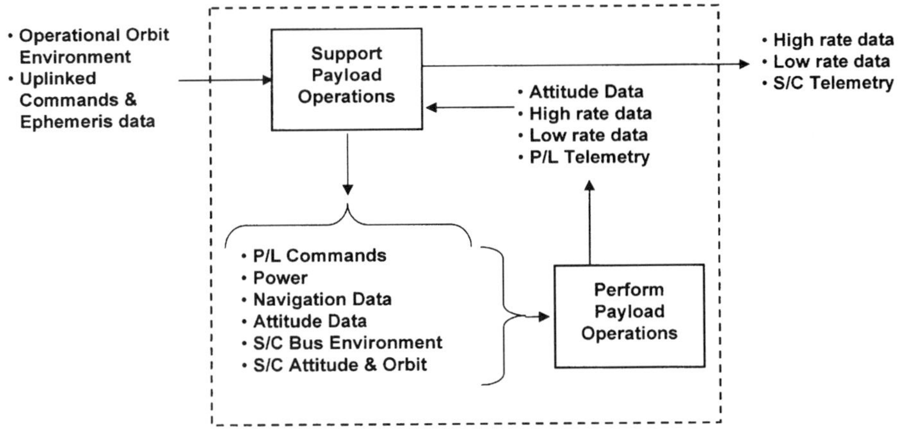

Figure 5.10 "Perform Satellite Operations" Function Decomposition.

Key Point

> This is a simple example, but the point is important — the implementation required at each level of the hierarchy may be very different for each phase of the life of the system. Therefore, system functions must be defined for each mission phase.

As discussed in the preceding chapters, requirements come first, then functions are identified from those requirements, and finally implementations are developed that provide the necessary functionality at the required performance level. By the way the customer defined the initial requirements set, it is apparent that he or she has conceptualized a design in which a spacecraft bus will support the telescope. It is this knowledge about how the system is to be implemented that enables the decomposition of the "Perform Satellite Operations" function into two sub-functions: "Perform Payload Operations" and "Support Payload Operations." The latter function, of course, is implemented by the spacecraft bus, which is the focus of this example development activity. This decomposition is depicted in Figure 5.10.

Notice that the inputs (operational orbit environment, and uplinked commands and ephemeris data) and outputs (high rate data, low rate data, and S/C telemetry) defined for the "Perform Satellite Operations" function in Figure 5.9 are still present in its decomposition depicted in Figure 5.10. However, they are applied more specifically to the "Support Payload Operations" function in the decomposition.

Figure 5.10 also shows the development of the interfaces between the two functions identified in the decomposition. As defined in the requirements in Table 5.1, the payload requires commands, electrical power, navigation data, attitude data, a controlled physical environment, and a particular orbit

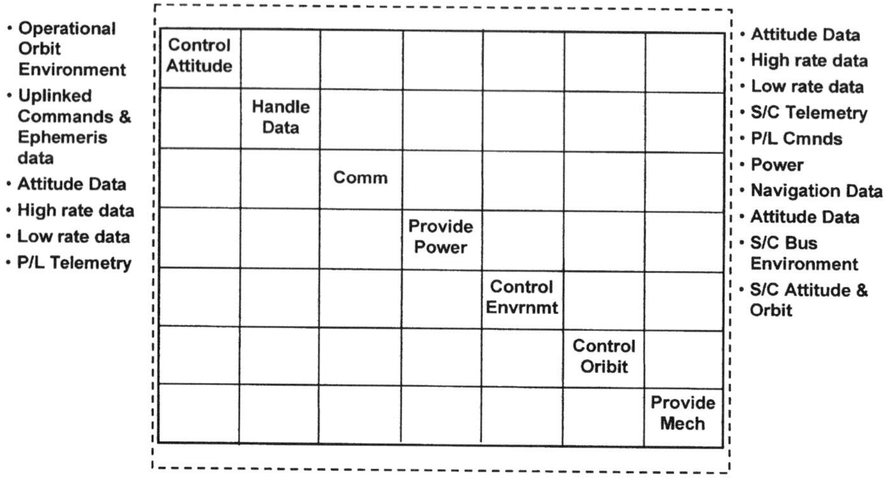

Figure 5.11 Support Payload Operations Decomposition.

and attitude within that orbit. In addition, the telescope must provide data to the spacecraft bus. Thus attitude data, high and low rate data, and telescope telemetry must be provided to the spacecraft bus.

As the decomposition of the "Support Payload Operations" function shown in Figure 5.11 indicates, the spacecraft bus must provide functions that perform the tasks required by the requirements set. Applying the same N^2 methodology again, each function is placed along the diagonal of the matrix. The matrix guides the design team to determine which, if any, interfaces are required between the various functions included in the decomposition.

Figure 5.12 illustrates the continuity between the various levels of decomposition. Also indicated is the implementation assumed that enabled each decomposition. The design assumption that enabled the decomposition of the "Perform Mission Operations" function was the concept of a separate spacecraft bus. This enabled a decomposition to two main functions at the next level down: "Support Payload Operations" and "Perform Payload Operations." The "Support Payload Operations" function, which is the spacecraft bus itself, was decomposed by assuming that the standard spacecraft bus subsystems would be implemented. These functions are identified along the diagonal, according to the N^2 format.

Operations Concept — The generation of the Operations Concept is initiated during the development of the functional description of the system. It typically describes how this system fits within the overall program in terms of specific capabilities and functions provided. It discusses management of the system during all operational modes for each mission phase and how the data is handled and generated by the system. This would include processing, storage, and distribution of the data. Other issues such as program organization, specific types of personnel and job functions necessary, equipment, and training are also addressed.

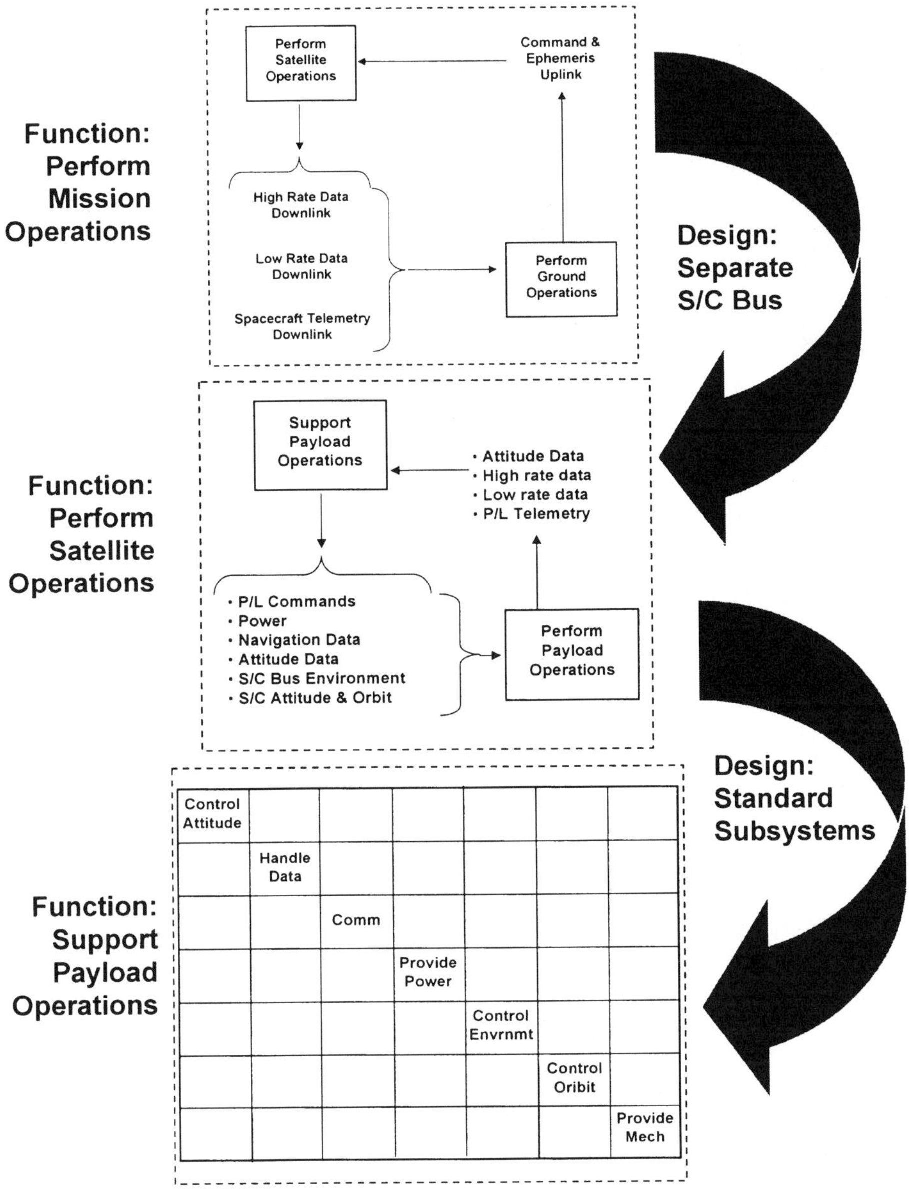

Figure 5.12 Decomposition Continuity.

Output

Specification(s)
Functional models (block diagrams, flow diagrams, behavior diagrams, simulations)

3. *Digression: Why Functional Analysis?*

Before proceeding to the discussion concerning Rework Discovery activities, the question "Why spend precious program resources performing functional analysis and functional decomposition?" is addressed.

Competitiveness — One important reason is competitiveness. A customer is generally more concerned with obtaining the functionality and performance levels desired than with the way in which that functionality is implemented. This, of course, assumes all else being equal, such as reliability issues. If a company is to remain competitive in an environment where technology is rapidly changing, it must focus on the functionality it is providing to its customers and not become overly enamored with its particular implementation or design. As an example, consider the slide rule. Since its invention in the early 1600s, it remained an important tool in the hands of scientists and engineers even well into the "space age." Many a slide rule was used in the design of the Space Shuttle. It was pervasive into the 1980s, but by the 1990s slide rules were little more than collector's items. What happened? More to the point of this discussion, what happened to the companies that manufactured them?

The slide rule was a calculator. Users purchased them for their functionality — performing mathematical calculations. Users were not so much concerned with the quality of the ivory and the fineness of the scales as they were with being able to perform calculations quickly, easily, and accurately. This was proven when the electronic calculator came on the scene. Within about a decade slide rules were a thing of the past. How many companies engaged in the manufacture of slide rules jumped into the electronic calculator market? Could it be that if those companies had understood their core competency as providing the functionality needed to perform mathematical calculations, they would have vigorously pursued technologies that better perform those functions? This is the danger with thinking primarily in terms of a particular design or implementation. An organization can become so consumed with its own method of providing a particular set of functions that it cannot leverage new technologies that might better perform that functionality. This leaves such a company vulnerable to competition that might be more agile. How much more is this true in those markets where technology is changing and advancing at unprecedented rates?

Specification Development — Specifications should focus on functionality and performance, not implementation. This gives the experts, those engineers receiving and responding to the specification, the flexibility to design an optimal solution. Presumably they understand the pertinent technologies and are therefore in the best position to develop an optimal design.

Exploitation of New Tools — Also, there are an increasing number of tools available to the designer to simulate functionality and performance. Among other things, this provides a means for validating a specification —

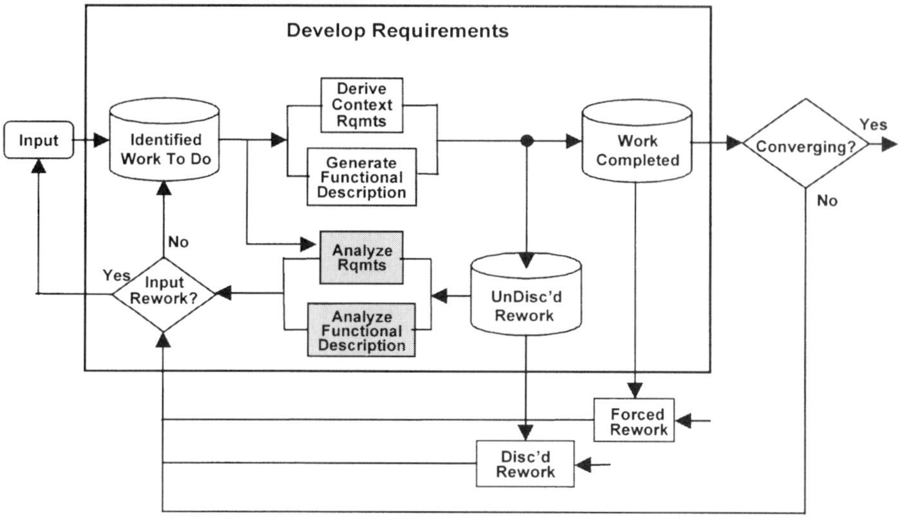

Figure 5.13 "Develop Requirements" Rework Discovery Activities.

ensuring that it is self-consistent and complete before committing design resources to what could be faulty input data.

Focus Research and Development (R&D) Efforts — Finally, if functionality and performance can be identified for future systems, research and development (R&D) efforts can be directed toward developing designs that provide them. In this way, direction can be given to R&D efforts in terms of identifying where resources should be directed in the development of new core competencies with the highest leverage for the organization.

C. *Rework Discovery Activities*

Figure 5.13 highlights the Rework Discovery activities that are performed within the Develop Requirements activity.

1. *Analyze Requirements*

This activity determines the validity of both the imposed and derived requirements. The goal is to ensure that the requirements are complete, self-consistent, unambiguous, and verifiable or measurable. In addition, it must be determined how the requirements will be verified and at what level verification will take place in the system build-up.

The task of analyzing the requirements set for problems is obviously a Rework Discovery activity. In terms of the logic of the overall SDF, Rework Discovery activities follow Work Generation activities — data must be generated before it can be analyzed for potential rework. Nevertheless, this task should be performed whenever new or changed requirements are input to the development activity. This is illustrated in Figure 5.13 by the arrow from the "identified work to do" bucket, indicating work is flowing not only to the Work Generation activities, but also directly to the Requirements Analysis activity. Because the sooner rework is discovered the less its potential impact to the program, input requirements are analyzed as soon as they are introduced.

Output

- Identification of all "To Be Determined" (TBD) holes in the requirements, with a closure plan
- Identification of conflicting or inconsistent requirements, with a closure plan
- Interpretation of vague or ambiguous requirements in order to review them with the customer and gain consensus
- Determination of the verification method (test, analysis, demonstration, simulation, inspection) that will be used for each requirement
- Determination of where in the system build-up each requirement will be verified
- Implementation of Configuration Management activities

2. Analyze Functional Description

The main task is to develop and validate the functional description of the system before resources are spent developing the design. This may involve simulation of the identified functions with their respective interfaces in order to verity that the specification is valid in terms of self-consistency. The main activities here include:

- Determination if the specifications are complete and self-consistent
- Identification of all functional requirements flowing out of imposed and derived requirements
- Determination of performance requirements of each function and the relationships (interfaces, interdependencies, etc.) between functions

Output

Validated specification(s)
Functional models (block diagrams, flow diagrams, behavior diagrams, simulations)

Key Point

> It should be noted here specifically that functional decomposition does not occur in the Requirements Development activity. A function cannot be decomposed

> without some knowledge and/or assumption regarding how it might be implemented.[43]

In the preceding discussion, several decompositions were developed. The first showed the entire ESAT system which included the space segment, ground segment, and launch segment, then the system was decomposed down to the spacecraft bus level. During the process, each assumed implementation that facilitated each decomposition was pointed out. In this way, the principle that functional decomposition cannot be performed apart from some knowledge or assumption about the implementation was reinforced. This discussion was presented in the context of the Requirements Development activity in order to show how the customer-imposed requirements for the spacecraft bus were derived by the customer before the Spacecraft Bus Development activity was initiated. This in no way implies that functional decomposition is performed in the Requirements Development activity. Rather, this discussion sought to emphasize the fact that functional decomposition is performed under the assumption of a particular implementation — which is developed in the Synthesis activity. Functional decomposition follows definition of the "how," which is developed in the Synthesis activity.

Other authors assert similar ideas. For example, Hatley and Pirbhai conclude:

> [H]igher levels of the system always provide the requirements for the lower levels. For systems containing hardware and software, this means that we need to know system-level requirements, decide on the system-level architecture, and then decide on the allocation of system requirements to hardware and software before we can establish the software requirements.[44]

In another portion of the book discussing similar issues, they note:

> This leveled repetition of functional requirements definition, followed by physical allocation, is fundamental to the nature of large systems development.[45]

[43] The author does not offer a formal proof of this sometimes debated point. However, to argue from practical experience, this author is not aware of any credible example of a functional decomposition, of either hardware or software, that has been performed without some reference to a design or design concept.

[44] Hatley, Derek J. and Imtiaz A. Pirbhai, *Strategies For Real Time System Specification*, New York: Dorset House, 1988, p. 264.

[45] Ibid., p. 7.

Similarly, Professor Nam Suh states:

> There are two very important facts about design and the design process, which should be recognized by all designers:
>
> 1. FRs [Functional Requirements] and DPs [Design Parameters, i.e., implementation] have hierarchies, and they can be decomposed.
> 2. FRs at the ith level cannot be decomposed into the next level of the FR hierarchy without first going over to the physical domain and developing a solution that satisfies the ith level FRs with all the corresponding DPs. That is, we have to travel back and forth between the functional domain and the physical domain in developing the FR and DP hierarchies.[46]

Functions and their respective implementation are intimately intertwined and necessarily dependent. Therefore, it is not possible to specify requirements that are independent from implementation in any absolute sense.[47] This has implications regarding the development of specifications in a multileveled hierarchy.

Key Point

- First, a specification is directly coupled to the implementation at the level above. It is therefore not implementation independent and cannot be so. It is incorrect to hold the notion that a specification can be written with no reference to implementation.
- Second, the System Development process cannot replace, nor is it intended to replace, technical expertise. In fact, because of the necessary connection between requirements and implementation, the SDF cannot be effectively applied without significant technical expertise.
- Third, a change in the functional description above will likely necessitate a change in the dependent implementation adjacent to it. Likewise, a change in the implementation above will likely necessitate a change in the dependent functional description of the system(s) below it (cf. Figure 5.30).

Output → Functional Description

[46] Suh, Nam P., *The Principles of Design*, New York: Oxford University Press, 1990, p. 36.
[47] This is in contrast to those who emphasize the necessity of specifying requirements in such a way that they are independent from implementation.

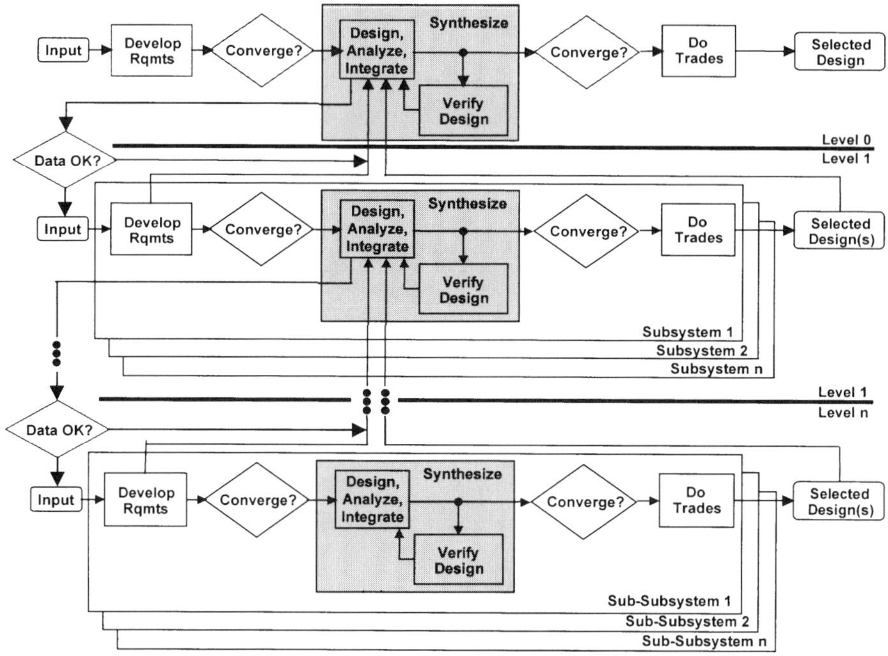

Figure 5.14 The "Synthesize" Activity in the System Hierarchy.

Customer Consensus — Although customer consensus is critical throughout the system development, because requirements drive the system design, concurrence from the customer community is especially crucial and is therefore specially noted here as an essential component to the Requirements Development activity.

II. Synthesis

The etymology of the word "synthesis" is συν + τιθεναι, "together with" + "to place". Synthesis means, therefore, "to place together with."[48] Synthesis has to do with the integrating of the elements of the solution into a coherent whole. Therefore, subsumed under this primary activity are all the subactivities involved in designing, analyzing, and verifying the system implementation. Figure 5.14 highlights the "Synthesize" activity in the context of the system hierarchy.

Figure 5.15 illustrates the decomposition of the Synthesize activity derived in Chapter 2. It comprises the Work Generation activity "Design, Analyze, and Integrate," and the Rework Discovery activity "Verify Design."

[48] *Merriam-Webster's Collegiate Dictionary, Tenth Edition*, Springfield MA: Merriam-Webster, 1996, p. 1197.

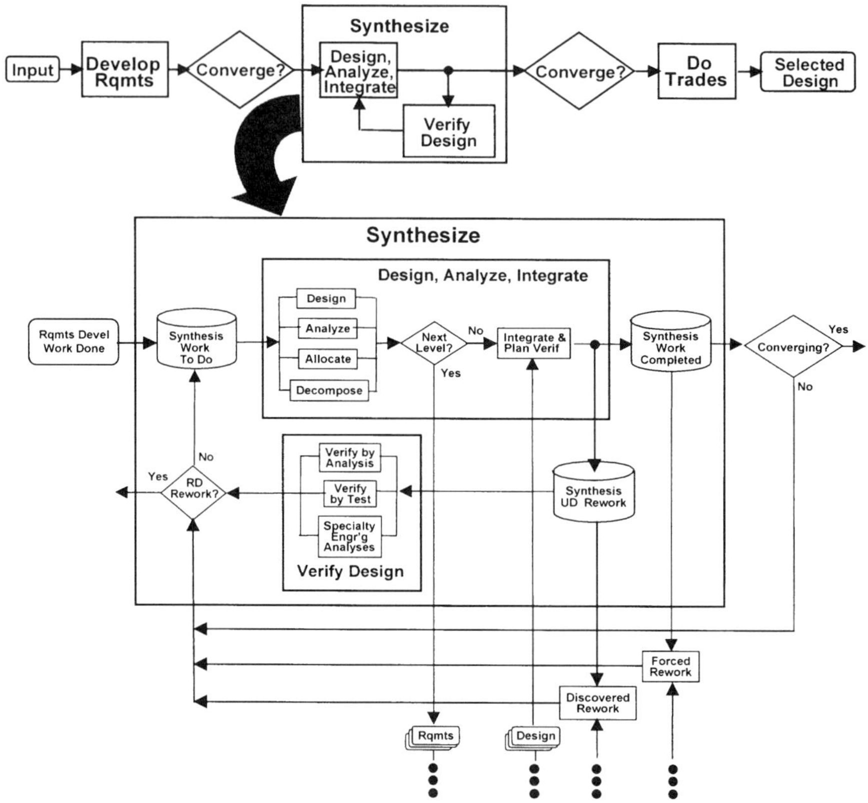

Figure 5.15 The "Synthesize" Activity Decomposed.

As mentioned previously, the System Development activity considers not only the development of the deliverable product itself, but also all the associated hardware, software, procedures, processes, etc. needed to produce, integrate, test, deploy, operate, support, and dispose of the system.

A. Work Generation Activities: Design and Integration

Determine "How" to Implement the "What" — Figure 5.16 highlights the "Work Generation" activities performed within the Synthesize activity. The activities identified are focused on generating the outputs needed at a particular point on the program timeline. As the program moves forward, the focus shifts from parametric analyses aimed at defining the available design space to detailed solutions in response to the increasingly detailed requirements.

1. Design

The Design activity is narrowly defined here as the set of activities that defines the initial design space, develops concepts or solutions in response

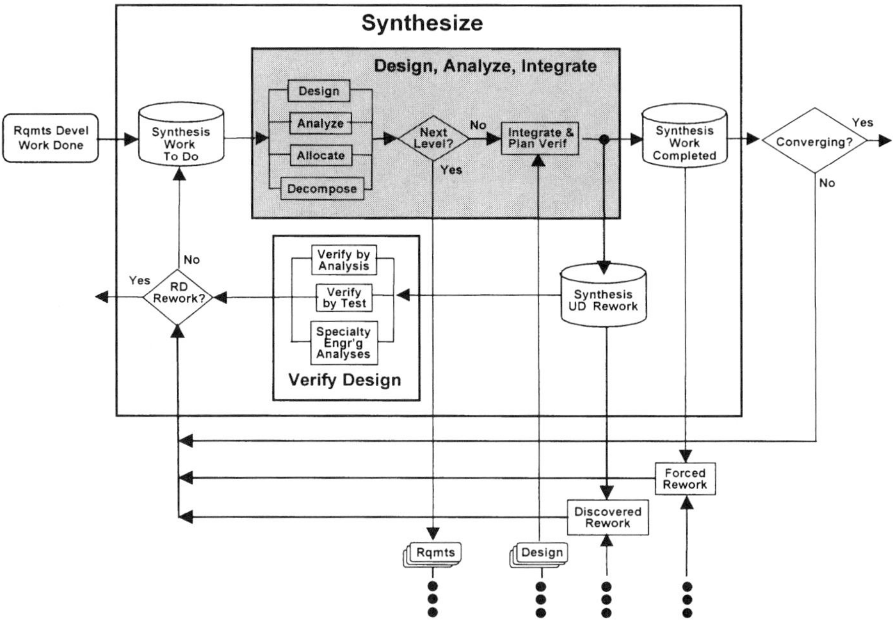

Figure 5.16 The "Synthesize" Work Generation Activities.

to the requirements, generates the design documentation, and performs analyses needed in the development of the solution.

- Quantify Design Space (H/W and S/W)
  - Parametric analyses
  - New technologies and heritage designs are surveyed for applicability
- Generate Preliminary and Detail Design
  - Block diagrams, schematics, drawings, etc.
  - Internal ICDs
- Risk Management → Identify and Assess Risk
  - Technical performance, cost, schedule
  - Preliminary mitigation approaches
- Configuration management of all design documentation

Output → H/W & S/W concept(s) and/or design(s), risk assessment

As indicated above, the first activity performed in the Design activity is to quantify the design space: That is, to define the major system interfaces and environments for each mission phase as well as to quantify critical design parameters in order to understand the cause-and-effect relationships between them. Figure 5.17 defines the top-level implementation of the ESAT system for the Launch and Orbit Acquisition Phase. The three top-level

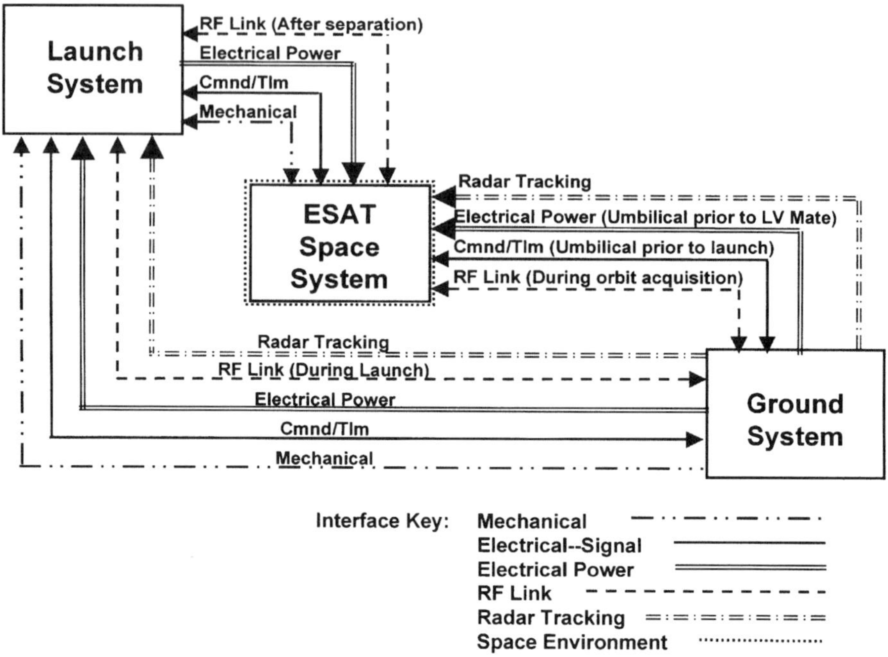

Figure 5.17 Interfaces — Launch and Orbit Acquisition Phase.

elements of the system are now described as nouns to indicate implementation, instead of verbs which are used to signify functionality. Thus there is direct traceability from the top-level functions to the top-level design or implementation. The major system elements are the ESAT space system, the ground system, and, during the Launch and Orbit Acquisition Phase, the launch system. Also included in Figure 5.17 are the interfaces between the system elements.

Figures 5.18 through 5.26 illustrate some of the various top-level architectures that have been implemented in the past and present, which could be employed as potential solutions for the ESAT spacecraft bus concept. For the ESAT example, the focus will be on the Attitude Determination and Control Subsystem (ADACS) to illustrate functional decomposition to the subsystem level. The attitude control requirements for a spacecraft play a significant role in determining many aspects of the bus design. The figures depict five different spacecraft types — all driven primarily by ADACS requirements. Some spacecraft missions require no attitude control. The Environmental Research Satellite (ERS), illustrated in Figure 5.18, was just such a spacecraft.

Some missions require that only one axis of the spacecraft be pointed toward the earth to a fairly loose tolerance. This can be accomplished by a "gravity gradient" design. The GEOSAT spacecraft, shown in Figure 5.19, took advantage of this concept. It was designed to measure sea surface heights.

Figure 5.18 The Environmental Research Satellite. (Photo Courtesy TRW, Inc.)

Other missions require a single axis be pointed toward the earth, but with a higher degree of accuracy. Spin-stabilization is often employed in these situations. The DSP (Defense Support Program) spacecraft — a military early warning system — is one example. Another example is the Television Infrared Observation Satellite (TIROS) II,[49] a meteorological satellite. These are depicted in Figures 5.20 and 5.21 respectively.

For those missions where all three axes must be stabilized, there are two primary designs: bias momentum and zero momentum. The bias momentum design is often used in geo-synchronous missions. One example is NASA's Tracking and Data Relay Satellite System (TDRSS), shown in Figure 5.22.

Especially in low earth orbit remote sensing missions, where a high degree of pointing accuracy is required, the zero-momentum design is often implemented. The Compton Gamma Ray Observatory (CGRO) and the Hubble Space Telescope (HST), shown in Figures 5.23 and 5.24, respectively, are examples.

Figures 5.18 through 5.24 illustrate an important aspect of system design: a thorough examination of existing designs to see if anything already developed might be suitable. This is a good approach because there is inherently less risk in designs that have already been proven, not to mention that the

[49] Newer TIROS satellites are three-axis stabilized, zero-momentum systems.

Figure 5.19 The GEOSAT Spacecraft. (Photo Courtesy Applied Physics Lab, Johns Hopkins University.)

cost to adapt an existing design is often less than developing a whole new concept. Of course, it is not always the case that an existing design is directly applicable. In such circumstances, the development of hybrids from existing designs might be appropriate. Other situations might necessitate that the design team start with a "clean sheet of paper" and develop a new concept from scratch.

In the early stages of development, parametric analyses are often performed. This is helpful where certain parameters are coupled in such a way that increasing or improving one parameter may have adverse effects on one or more other parameters. These kinds of analyses show how changes ripple through the design in terms of their effects on other parameters.

For example, for a mission in which high resolution images will be taken, there will be trade-offs between the resolution (sharpness) of the images taken and the footprint or coverage of each image (an entire geographic region or a small city block). This trade-off must also consider on-board data storage and/or downloading the data to ground stations. Depending upon the downlink, a fixed amount of data can be transmitted to the ground at any given opportunity for ground contact. This will limit the amount of data that should be generated by the telescope. The trade-off then becomes one

Figure 5.20 The Defense Support Program (DSP) Spacecraft. (Photo Courtesy TRW, Inc.)

of resolution vs. image size. If the downlink capacity is too restrictive, this analysis may indicate a trade-off of data quality vs. the cost of adding more downlink capacity. This is just one simple example of what sort of parametric analyses can be performed to quantify design space.

2. *Analysis*

Figure 5.25 highlights the Analysis and Allocation activities in the context of the "Design, Analyze, Integrate" activity. The "Analysis" activity includes any and all analyses necessary to support quantification of design space and design parameters, as well as to ascertain technical, cost, schedule, and risk performance of the system. Since the goal of this book is simply to establish a framework for complex system design, it is beyond the scope of this discussion to elaborate on these. Therefore, only some of the myriad analyses that might take place on a given system development program are listed:

- Mission, system, electrical, digital, analog, RF, mechanical, etc.
- Simulations
- FMECA (Failure Modes Effects and Criticality Analysis)

Figure 5.21 TIROS II Spacecraft. (Photo Courtesy NASA.)

3. *Allocation*

Allocation involves not only technical elements, but also the cost and schedule components of the system development. There are also several managerial activities that are associated with the allocation activity. Thus, allocation involves the following:

- Allocate functionality, performance, constraints to H/W and S/W elements
- Define budgets
  - Technical: mass, power, throughput, memory, RF links, etc.
  - Reliability, contamination, etc.
  - Margin and contingency rules
- Configuration management — Controlling the budgets
- Risk management — Assessing convergence, as per Figure 5.26
- Performance monitoring, metrics development, defining/refining TPMs
- Cost and schedule management

Figure 5.22 Tracking and Data Relay Satellite (TDRS). (Photo Courtesy TRW, Inc.)

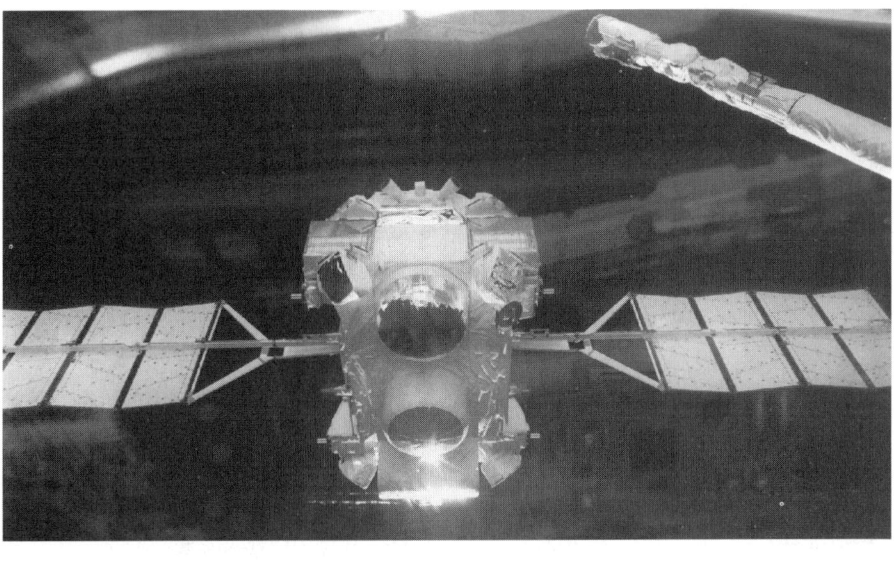

Figure 5.23 The Compton Gamma Ray Observatory (CGRO). (Photo Courtesy TRW, Inc.)

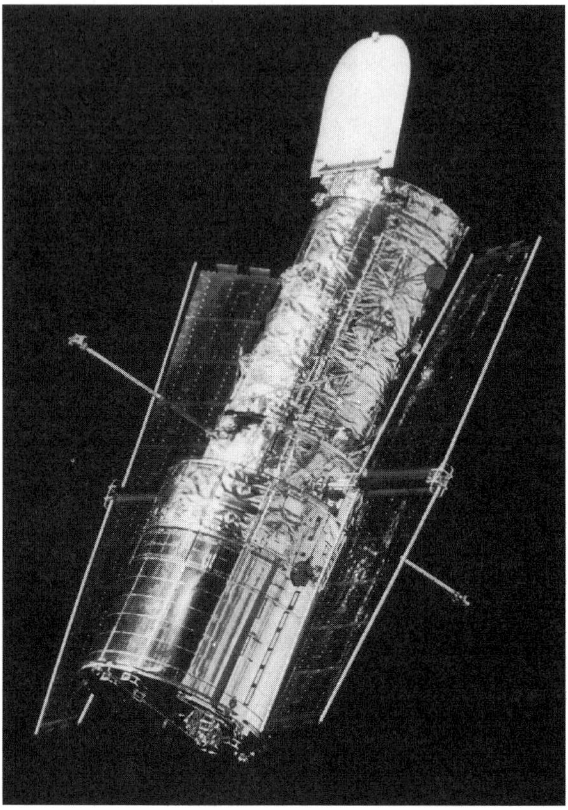

Figure 5.24 Hubble Space Telescope (HST). (Photo Courtesy NASA.)

Some important questions arise in any discussion concerning margin and its role in the allocated budgets: How much margin should be included? How should this change over time? Figure 5.26 provides a notional depiction of how margin should converge to a small percentage of the budget as uncertainty in the design decreases. The point of the figure is not to prescribe specific numbers, but rather to suggest that *margin must be factored into the development activity and monitored to ensure that the system is converging upon a low-risk solution*. The upper and lower curves represent reasonable amounts of uncertainty as the development progresses. If these bounds are exceeded, unforeseen risk may be indicated.

Output → Budgets, technical performance measures

Figure 5.27 illustrates how functionality is derived directly from the input requirements and that the implementation is driven by the functionality and associated performance required. Defining functionality without defining the required performance is not useful to the designer. As discussed

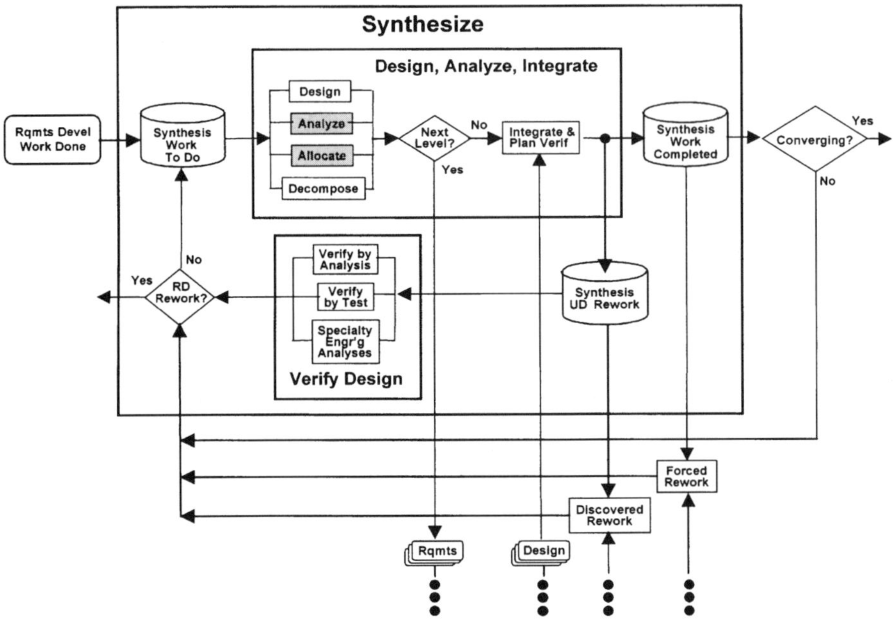

Figure 5.25 The "Analyze" and "Allocate" Activities.

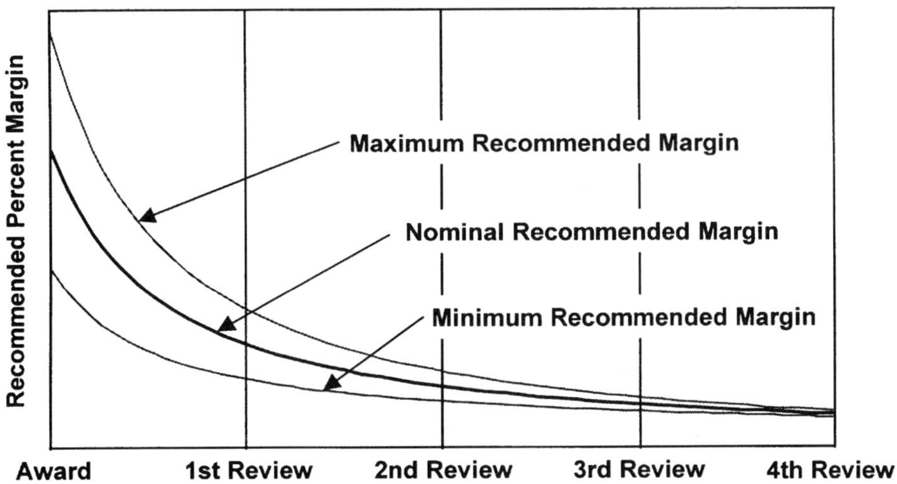

Figure 5.26 Notional Convergence of Margin and Reduction in Uncertainty.

above, the ADACS implementation chosen is based primarily upon the accuracy the subsystem must provide to the spacecraft. Some spacecraft require no ADACS system because the payload is not dependent upon pointing the satellite in any particular direction.

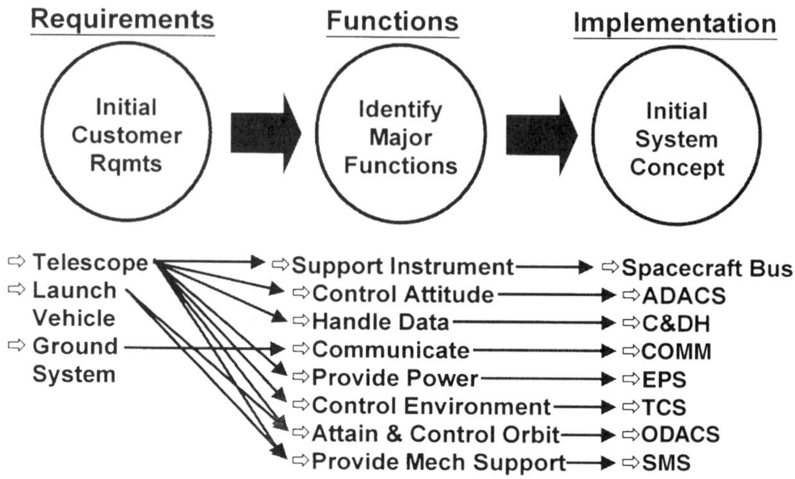

Figure 5.27 Allocation of Functionality to Implementation.

Key Point

> The point here is that there must be traceability between requirements, functionality, and implementation. Functions must be allocated to specific elements of the design. This ensures that the design is appropriate for the intended use. Resources are not wasted by over-design and the mission is not unsuccessful because of insufficient capability.

If a certain system element has no functionality allocated to it, the question ought to be raised as to why that element is included in the system. Of course, where heritage designs are used, it may be cost and schedule effective to retain functionality that is not required, simply because it is cheaper and/or more schedule efficient not to eliminate it from the existing design.

As shown in Figure 5.27, seven subsystems have been identified for ESAT: attitude determination and control system (ADACS), orbit determination and control system (ODACS), command and data handling (C&DH) subsystem, electrical power subsystem (EPS), communications (COM), propulsion subsystem (PRS), structures and mechanisms subsystem (SMS), and the thermal control subsystem (TCS).[50]

[50] Software is not identified as a separate subsystem, but is included in the implementation of each appropriate system element. The spacecraft bus as a system includes both hardware and software, as do most or all of the major subsystems. Allocation of a function or set of functions to hardware or software is a decision made during the design development. While software may be developed by a distinct functional organization, it is not viewed herein as a separate subsystem in terms of the design itself. System-level software is managed at the system level of the design; subsystem-level software is managed at the subsystem-level of the design; and so on. When a particular subsystem is discussed, it is assumed that it is comprised of all its constituent elements, including both hardware and software.

At the spacecraft bus hierarchical level, technical budgets are generated for each identified subsystem. The technical budgets define mass, electrical power, memory, throughput, etc. for each subsystem. These budgets are very important because, at the early stages of the development, they can be used to determine the risk level of the program to a significant degree. Early in the program, the design effort focuses on the upper levels of the design. Often, budgets are allocated to lower-level system elements without detailed analyses to validate the budgets. There is, therefore, risk introduced into the program because there is some probability that those system elements cannot be accommodated within their assigned budgets. An element may require more allocation of mass, power, or other resources. Hopefully, such a problem can be accommodated by reallocation of margin or contingency. If not, the ripple effects can be significant.

Figure 5.28 depicts a matrix where each allocated resource is represented as a column and each subsystem as a row. The figure illustrates only mass and power. However, resource budgets should include not only these two, but also all others as well. Other resources might include: memory, throughput, communication links, etc. The matrix should include the current value of the resource used by the element, the current budget, and the margin remaining for each. Risk areas can be identified based upon the rate at which a particular resource is being consumed as the design matures.

4. *Functional Decomposition*

Figure 5.29 highlights the "Decompose" activity of the "Design, Analyze, Integrate" activity.

As discussed previously, a function cannot be decomposed without some knowledge of "how" the system will be implemented. Therefore, functional decomposition to the next level down in the hierarchy logically follows the generation of concepts at the level above. Thus, it is here in the overall process that functional decomposition is performed. The following series of activities are necessary to properly decompose a system or system element.

- Receive function and performance requirements from the Requirements Development activity
- Develop design concepts that implement the identified functions at the required performance levels
- Decompose the implementation into subfunctions for the next-level-down activity
- Identify the interfaces between the subfunctions
- Partition subfunctions into logical groups (potential subsystems); group the functions such that interfaces are minimized between logical groups
- Generate the functional model and verify the functional definition
- Generate function and performance requirements (specifications and ICDs) for each logical grouping of functions

	Mass				Power			
	Current	Budgeted	Margin		Current	Budgeted	Margin	
			Number	Percent			Number	Percent
Total Spacecraft	2170	2400	230	9.6%	953	1105	152	13.8%
Telescope	925	1000	75	7.5%	300	350	50	14.3%
Spacecraft Bus	1245	1400	155	11.1%	653	755	102	13.5%
Structure & Mechanisms	450	500	50	10.0%	8	10	2	20.0%
ADACS	85	100	15	15.0%	20	25	5	20.0%
Earth Sensor								
Star Tracker								
Sun Sensor								
RWAs								
IMU								
Torque Rods								
Etc.								
EPS	400	450	50	11.1%	400	450	50	11.1%
Batteries								
Solar Arrays								
Distribution								
Etc.								
Comm	65	75	10	13.3%	125	150	25	16.7%
Transmitters								
Antennas								
Etc.								
C&DH	45	50	5	10.0%	38	45	7	15.6%
Command Processor								
Data Storage								
Command Distribution								
Etc.								
Thermal Control	20	25	5	20.0%	55	65	10	15.4%
Thermal Blankets								
Heaters								
Thermisters								
Etc.								
Propulsion	180	200	20	10.0%	7	10	3	30.0%
Hydrazine Tanks								
Thrusters								
Valves								
Etc.								

Figure 5.28 Allocation of Technical Budgets.

- Release function and performance requirements to lower-level development activities
- Receive feedback from lower-level development activities and refine specifications and ICDs
- Iterate as necessary

In Figure 5.30, L0 indicates Level 0 which, for this example, represents the top level of the system hierarchy. L1 indicates the next level down, or the subsystem level of the hierarchy. Notice that there are three subsystems indicated in the figure at level L1; often there are many more. Again, making use of the above definition of "system," it is asserted that this approach to functional decomposition is applicable to all levels of the system hierarchy. First, imposed requirements are input to the Requirements Development activity. Next, the functions with their performance requirements are analyzed and

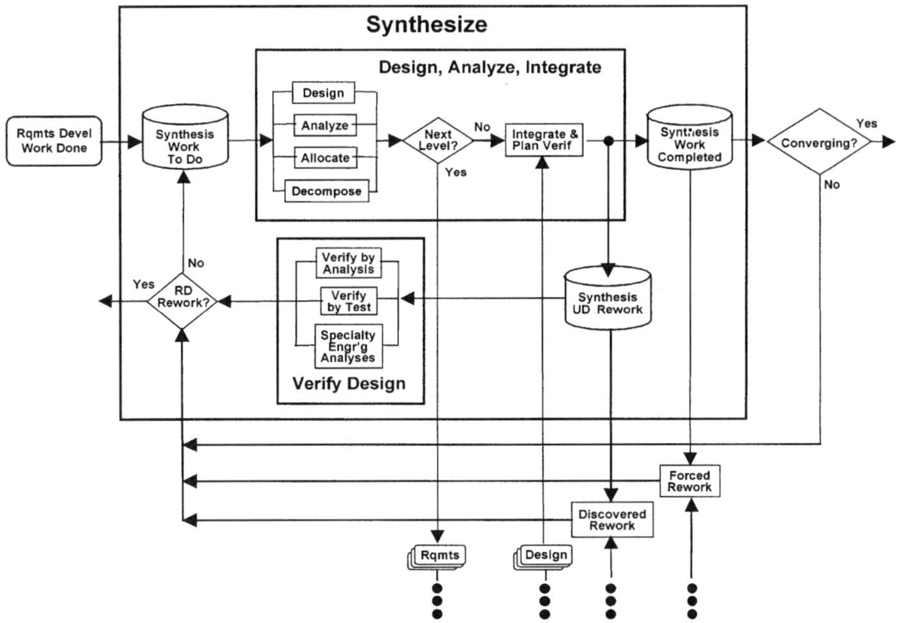

Figure 5.29 The Decompose Activity.

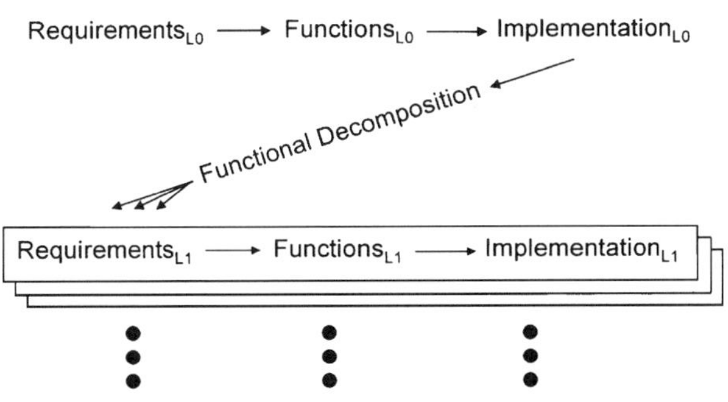

Figure 5.30 Functional Decomposition Methodology.

coalesced into a functional model and input to the Synthesis activity. Then, implementations are developed in response to the functional model. Finally, with a concept in mind, functional decomposition is performed as next-level-down functions are derived from the implementation candidate. These functions and required performance parameters are then collated into a requirements set as input to the next-level-down Requirements Development activity and the cycle is repeated as necessary.

Figure 5.31 illustrates the organic connection between implementation and functionality. Within the same level of the hierarchy, the requirements set (or functional architecture) drives the implementation architecture. Between tiers, the implementation at the tier above drives the functional

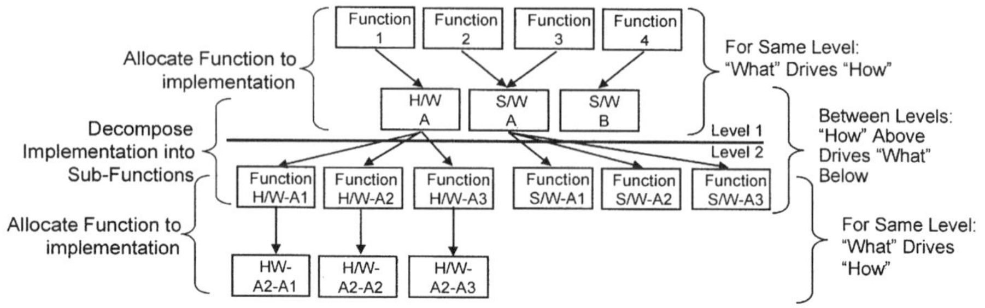

Figure 5.31 The "How" and "What" Relationship.

architecture of the subsystems below. Notice in the figure that more than one function can be allocated to the same design element. However, only one element implements any particular function. In other words, neglecting issues such as redundancy, the exact same function should not be provided by multiple components of the system. This could lead to confusion and cause malfunctioning of the system. This may not be the case, however, where the same function is needed in differing contexts.

One of the key issues involved in the decomposition of an implementation is the partitioning of the derived functions into subsystems. Pimmler and Eppinger note, "For a complex product . . . there are thousands of possible decompositions which may be considered. Each of these alternative decompositions defines a different set of integration challenges."[51] A key criterion to optimize the partitioning is the minimization of the number of interfaces or interdependencies between the identified functions in order to minimize the integration problem.[52]

Figure 5.32 illustrates an overview of the development by decomposition process, focusing on the attitude determination and control subsystem. The relevant input requirement originates with the customer's telescope. It requires that the spacecraft bus control the telescope attitude along all three axes to an accuracy of 0.01 degrees (requirement 2.2.8 from Table 5.1). Thus the function flowing from this requirement is "control instrument attitude." This function must be achieved at a performance level of 0.01 degrees along all three axes. This portion of the functional analysis is complete now that the function has been identified with its respective performance requirement. With the function identified, candidate designs can be generated. There are often many ways to implement a particular set of functions. In this simple example that is the case. There are several ADACS architectures that can be considered to implement the required functionality. Four potential design

[51] Pimmler, Thomas U. and Steven D. Eppinger, Integration Analysis of Product Decompositions, *Design Theory and Methodology*, DE-Vol. 68, ASME 1994, p. 343.

[52] The Design Structure Matrix (DSM) provides a useful tool for determining the best segregation of functional elements into collected sub-elements. Cf. Robert P. Smith, Steven D. Eppinger, "Identifying Controlling Features of Engineering Design Iteration," *Management Science*, vol. 43, no. 3, pp. 276-293, March 1997.

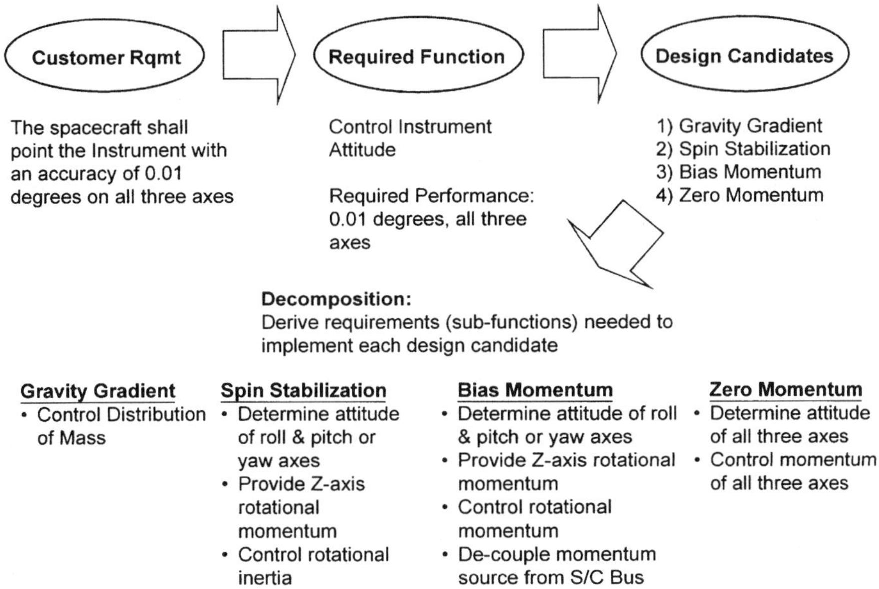

Figure 5.32 Second Level Decomposition.

solutions have been identified: gravity gradient, spin stabilization, bias momentum, and zero momentum.

While this is certainly not a book about spacecraft attitude control, it might be helpful to give a brief explanation of how each of these designs work.

Gravity Gradient — Gravity gradient uses the force of the earth's gravity field to cause the spacecraft to point one axis toward the earth. The mass of the spacecraft bus and payload components is typically concentrated at one end of the spacecraft (Figure 5.19). Usually a second mass, less than the first, is deployed a relatively large distance from the first. The force of gravity on the first mass is greater than that on the second and it is this "gradient" that causes the minor axis of the spacecraft to point in the proper direction within a few degrees. In general, there is no control of spin along the axis pointed toward the earth.

Spin Stabilization — Spin stabilization is implemented by spinning the spacecraft about the axis pointed toward the earth (Figures 5.20 and 5.21). The inertial force created by the spinning motion creates a stabilizing force along the spin axis, much like a spinning top or gyroscope. The downside of this control system is that the spacecraft, or a portion of it, must spin (some spacecraft implement a "despun platform" to offset this problem). This is adequate for certain missions but not for others.

Bias Momentum — Bias momentum uses the same inertial force as the spin stabilized spacecraft, but it is contained in a device called a momentum wheel (Figure 5.22). The momentum wheel spins, but the spacecraft itself does not. This allows all three axes to be stabilized. Momentum stored in the momentum wheel is managed by thrusters and/or torque rods that leverage the forces of the earth's magnetic field. One of the major drawbacks of spin stabilized and bias momentum stabilized systems is that maneuverability is difficult. A second drawback is that the spinning motion can induce disturbances (vibrations) into the payload.

Zero Momentum — Zero momentum systems also stabilize all three axes but without using any spinning motion for that purpose (Figures 5.23 and 5.24). The small disturbances encountered are taken out by storing the momentum created by those disturbances in devices called reaction wheels. Most zero momentum systems have one reaction wheel on each axis that is used to store momentum induced on that axis. Magnetic torque rods are used to "dump" momentum periodically. At opportune points in the orbit, torque rods are turned on that generate a magnetic dipole which interacts with the earth's magnetic field to apply the appropriate torque to the satellite. This torque allows the reaction wheel to slow down, or "dump," its momentum with minimal net disturbance to the spacecraft.

This explanation of spacecraft attitude control systems has been necessarily brief. The purpose is simply to facilitate an understanding of design by decomposition, using the spacecraft attitude control system as an example. If the reader requires a more extensive understanding of spacecraft attitude control systems, there are a number of excellent texts that can be consulted.[53]

Consider again the example illustrated in Figure 5.32. Once a subsystem architecture has been identified it can be decomposed into subfunctions. Notice that each ADACS architecture has a different decomposition since the functions required to implement each architecture are different. The gravity gradient architecture requires only that the mass of the spacecraft be distributed in a certain way; no other sensors or effectors are necessary. The spin stabilization and bias momentum architectures require similar functionality since both use rotational inertia created by a spinning mass and both require that the attitude of the roll and pitch or yaw axes be determined. The key difference between them, in terms of basic functionality, is that the bias momentum system requires that the source of the rotational inertia, the momentum wheel, be decoupled from the rest of the spacecraft. This decoupling allows the spacecraft to remain fixed relative to the earth while the rotating momentum wheel generates and maintains the stabilizing rotational inertia. Finally, the zero momentum architecture requires that the attitude of all three axes be determined and that the momentum of all three axes be maintained at zero.

At this point functional analysis is commenced by identifying the interfaces between the identified functions. The focus will be on the zero momentum architecture as this example is continued. Figure 5.33 represents the

[53] See, for example, Wertz, James R., Ed., *Spacecraft Attitude Determination and Control*, D. Reidel, Dordrecht, The Netherlands, 1997 reprint.

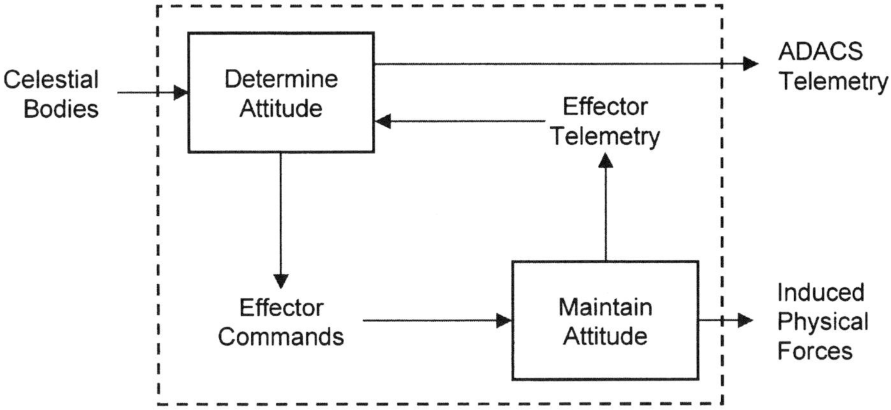

Figure 5.33 "Control Attitude" Function Decomposed.

decomposition of the zero momentum architecture. There are two primary functions necessary: determine attitude and maintain attitude. The figure depicts these two functions and their interfaces.

As has been emphasized, in order to perform decomposition there must be some understanding as to what the implementation will be.[54] As the decomposition of the function "Determine Attitude" is approached, the question arises, "how might this be implemented?" Several different methods for determining spacecraft attitude could be conceived: rate integration using gyroscopes, data from the Global Positioning System, uplink attitude data from the ground. For different reasons, these concepts may or may not be viable. The basic questions that must be considered here are "What resources are available?" and "What performance level is required?"

For ESAT, the focus will be determining attitude by exploiting the availability of celestial bodies. Now that a concept has been identified, a functional decomposition can be performed because the functions that must be performed in order to determine attitude using celestial bodies are now known. First, the celestial bodies must be sensed and the appropriate data must be generated, allowing the spacecraft to make use of that information. Second, the spacecraft needs to receive the data concerning the sensed celestial bodies, so that it can process it to determine its attitude. Finally, appropriate commands must be generated and distributed that will tell the spacecraft effectors what to do in order to maintain the desired attitude. As before, the interfaces between these functions must be identified and characterized. Figure 5.34 depicts the results of this process.

Figure 5.35 shows the functional decomposition of the "Maintain Attitude" function. In order to counter disturbances the spacecraft will experience, the system must be able to generate rotational momentum opposite

[54] To reiterate, this does not imply that great detail must be provided. The point is simply that at least a concept of how the function will be implemented is necessary before decomposition can commence.

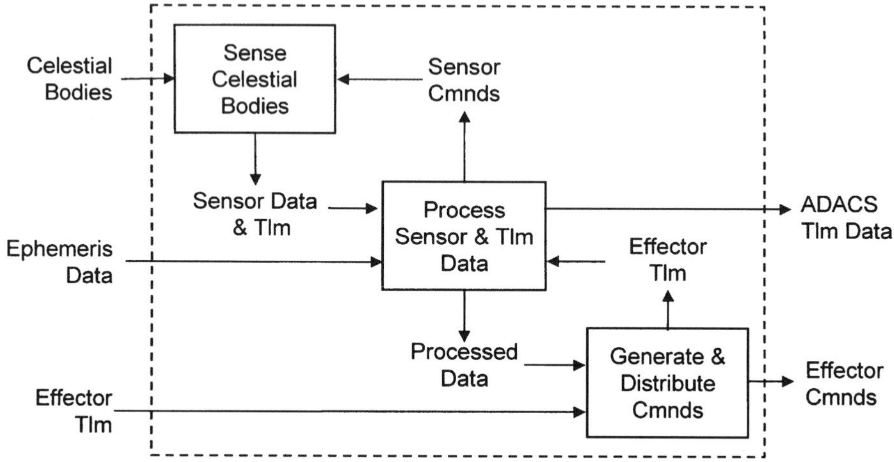

Figure 5.34 "Determine Attitude" Function Decomposed.

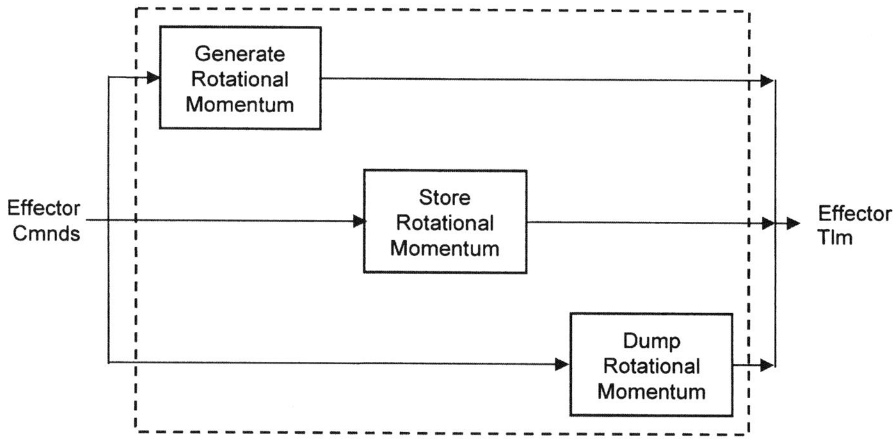

Figure 5.35 "Maintain Attitude" Function Decomposed.

the direction of the disturbance. Therefore, the "Generate Rotational Momentum" function is included in the system. Second, any momentum removed from the satellite must be stored until it can be dumped, thus a "Store Rotational Momentum" function is provided. Finally, because the rotational momentum storage devices will not have infinite capacity, a function is needed to "dump" rotational momentum periodically. The interfaces between these functions must also be identified and characterized.

The preceding discussion provides an example of one of the architecture candidates (zero momentum). If the other architectures were serious candidates, similar diagrams for each would be developed. The next step in the framework is to develop requirements for the next-level-down Requirements Development activity.

Output → Lower-level validated specifications and ICD(s), lower-level simulation

5. *Inter-Level Interface*

It is at this point in the SDF where the vertical interface between system elements occurs. This is the interface between the System and Subsystem activities, for example. This same interface occurs at all vertical interfaces.[55] Requirements are passed down the hierarchy and design data and other data are fed back here.

Information Flow-Down — Without elaboration (this is beyond the scope of this book) it is suggested that any "design-to" data must be configuration controlled to some degree. The formality and rigor of the configuration management effort must be commensurate with the program need.

Data Feedback — Development issues relating to reallocation and redesign must be managed. In order to reallocate effectively, knowledge of where margin and contingency reside in the system is necessary. Therefore, this information must be documented in the system budgets.

Output → Configuration controlled documentation

6. *Integration*

Figure 5.36 highlights the "Integrate and Plan Verification" activity that is performed within the "Design, Analyze, Integrate" activity.

The Integration activity is a key one. This is the point in the process at which all the elements of the design are integrated or synthesized into a coherent whole. A major concern of the integration task is interfacing between the various system elements. This task is primarily concerned with synthesizing the design by updating design data, and managing the activities identified previously and bulleted below.

- Identify and characterize interfaces
- Note specifications, ICDs, databases, etc.
- Update design definition
- Update mission timeline and operations concept
- Block diagrams, schematics, drawings, layouts
- Management activities
    - Performance Measurement — Budgets, etc.
    - Subcontract Management
    - Risk Management — Identification, assessment, and mitigation approaches
    - Configuration Management — Configuration Control Board (CCB)

Output → Integrated design that includes the data generated above

[55] This is also depicted in Chapter 6 in Figure 6.4, "Program Team Interactions."

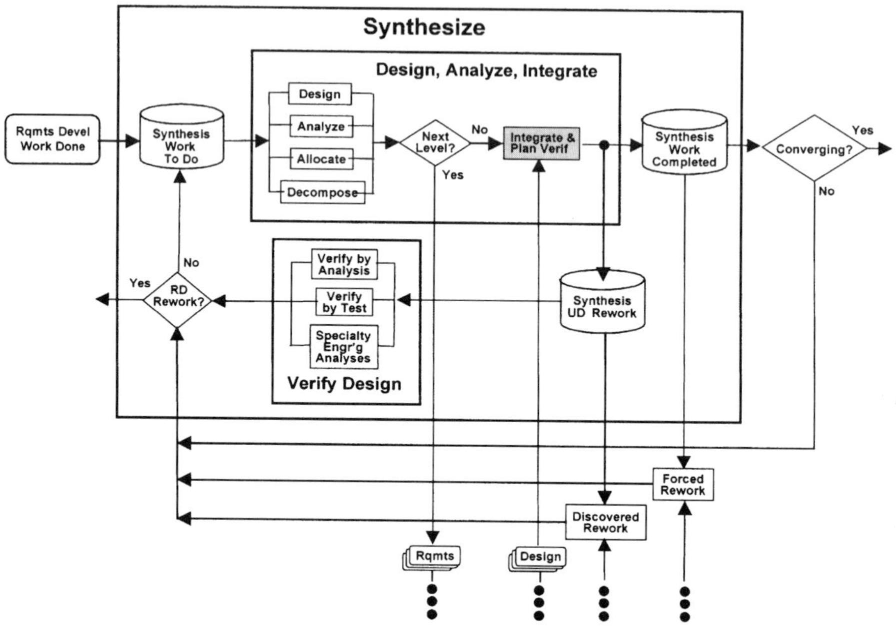

Figure 5.36 The "Integrate and Plan Verification" Activity.

Figure 5.37 provides a view of the integrated ESAT spacecraft in the form of a system block diagram, integrating the zero momentum ADACS architecture. Both the telescope and the spacecraft bus are shown, with key interfaces between them identified. The payload system is identified as such in the figure; all else represents the spacecraft bus. Each subsystem is shown with major components. Both the internal and external interfaces are identified.

B. *Rework Discovery Activities: Design Verification*

Assess "How-Well" — Figure 5.38 highlights the Rework Discovery activities performed within the Synthesize activity. The primary focus is verification of the developing design.

Verification involves two basic activities: Design Verification which focuses on the developing design, and Product Verification which focuses on the deployed system. Verification is accomplished by performing test, analysis, simulation, demonstration, or inspection. In this section, the primary concern is with Design Verification in order to show how it is involved in the development of the system design.

Key Point

> The Verification activity is an important part of the SDF. It is performed at the earliest stages of development and continues through the entire design phase.

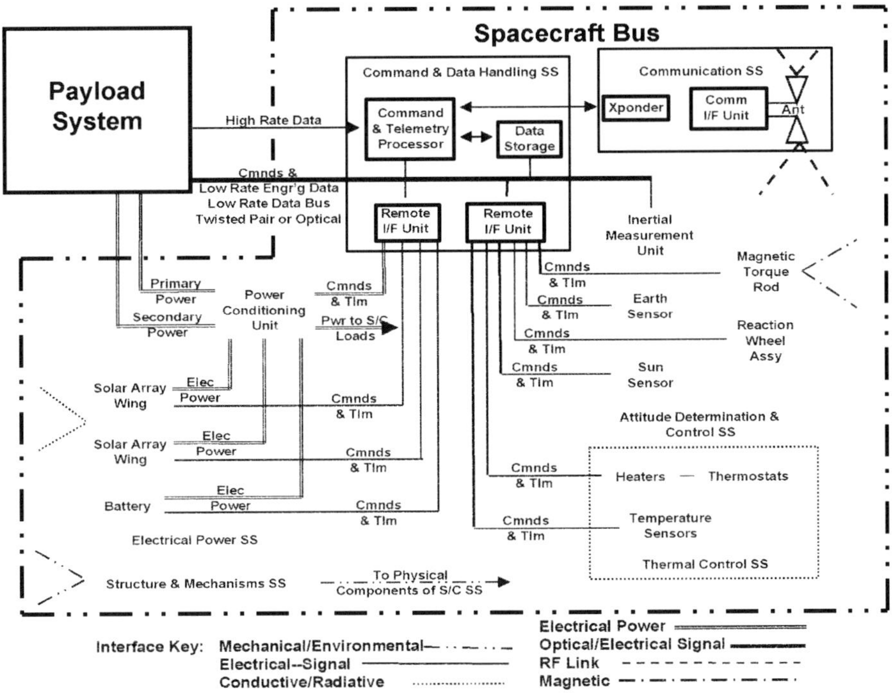

Figure 5.37 Integrated Spacecraft System: A Notional System Block Diagram.

1. *Analysis and Test*

a. *Analysis*

Those analyses aimed at determining "how well" the current design meets its requirements; these are in contrast to those analyses aimed at defining design space which are performed as described previously in the design activity.

b. *Test*

- Planning Activities (e.g., test requirements, test flow, resource planning, etc.)
- Testing Activities (e.g., engineering test models, prototypes, breadboards)

2. *Producibility, Testability, and Other Specialty Engineering Activities*

This activity assesses those areas of the design commonly called "specialty engineering" concerns.

- Is the design testable within resource and time constraints?
- Is the design producible within resource and time constraints?
- Is the design acceptable with respect to EMI/EMC, reliability, maintainability, affordability, supportability, etc. parameters?

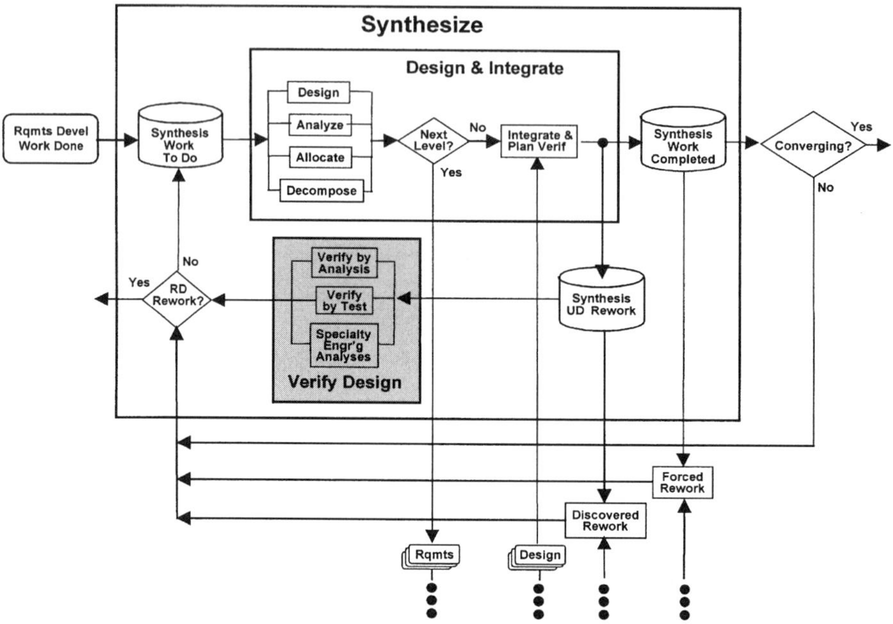

Figure 5.38 The "Synthesize" Rework Discovery Activities.

Output → The output of the Design Verification activity is a design that has been assessed as to how well it meets all the requirements.

III. Trade Analysis

Figure 5.39 illustrates where Trade Analyses are performed in terms of the logical sequencing of activities. Trades logically occur after Synthesis because the trade criteria (technical, cost, schedule, risk)[56] must be developed in order to make the selection.

This in no way implies that trade studies must wait until competing designs are complete. The reader is reminded that the present discussion takes place in the context of the Logical Domain, not the Time Domain. Only the logical sequencing of activities on the micro-time scale is in view here (recall Figure 3.1 and related discussion). Although generally emphasized in the early phases of development, trade analyses can occur during all development phases as the SDF is performed and iterated. The point here is that trade studies are dependent upon technical, cost, schedule, and risk criteria which can only be developed with reference to implementation. Absent such criteria, there is no basis for making a selection.

[56] Chestnut concurs, "Commonly accepted bases for judging the value of a system: (a) performance; (b) cost; (c) time; (d) reliability; (e) maintainability." Harold Chestnut, *System Engineering Tools*, New York: John Wiley & Sons, 1965, p. 11.

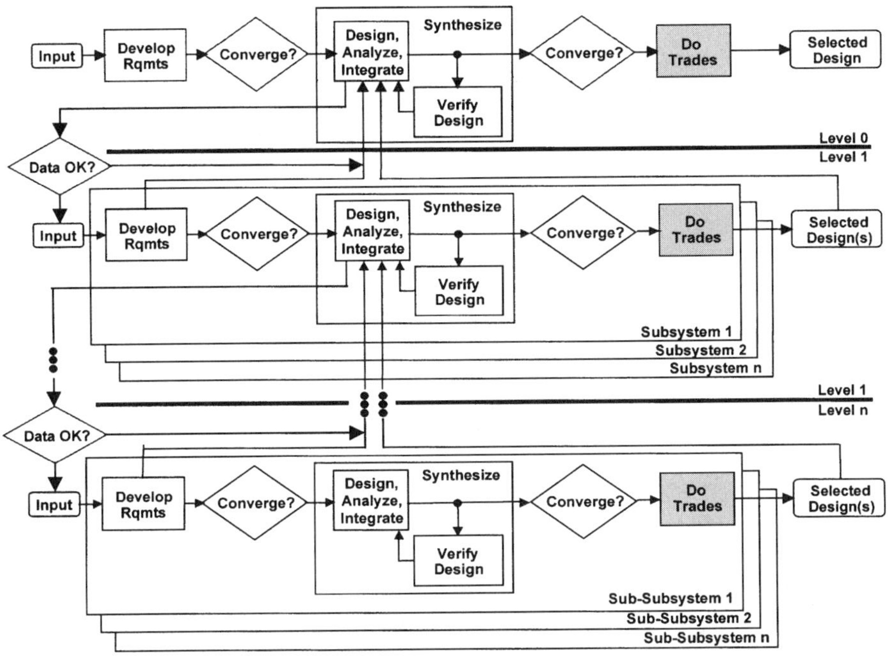

Figure 5.39 The "Do Trades" Activity.

There are many selection methodologies that can be employed in the selection process.[57] It is not the purpose here to discuss the pros and cons of the various trade methodologies found in the literature, but simply to delineate where in the System Development process Trade Analyses occur. The amount of rigor applied to the Selection process should be commensurate with customer requirements, program need, and other criteria as determined by the development team.

If multiple candidates emerge from the design and analysis activity, the selection process is implemented. However, if multiple candidates are compliant and equally acceptable to the design team, integrate each into the element of the level above and analyze for system benefit there. The selection is then made at that above level.

Here in the Trade Analysis section, it is appropriate to highlight the classic trade-off that occurs in most any System Development activity: cost, schedule, and technical performance. Issues relating to risk, robustness, safety, etc. could also be included, but certainly cost, schedule, and technical performance are central concerns. This is illustrated in Figure 5.40.

It is not often that all three of these issues can be improved simultaneously. In general, one or two can be improved at the price of the third or other two. The author once worked with an engineer who had a similar

[57] e.g., Pugh; DSMC; McCumber, William H., *System Performance Representation: Standard Scoring Functions*, NCOSE 1995, P003; Ulrich and Eppinger, pp. 105-122.

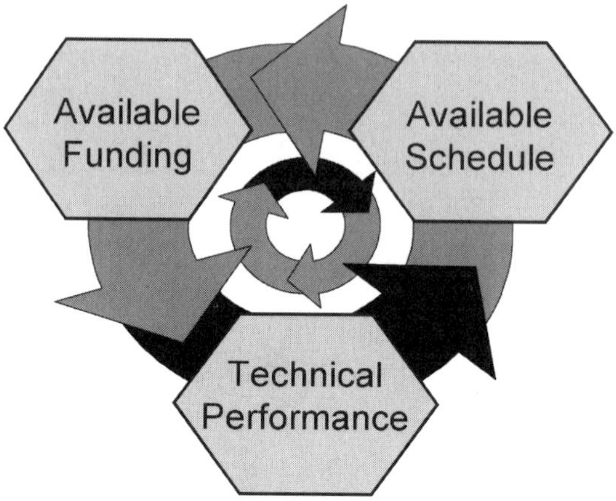

Figure 5.40 The Classic Trade-Off.

graphic in his office. It had the caption, "faster, better, cheaper — pick any two!" While this may not be an absolute truth in every situation, it does serve to emphasize the non-trivial trade-off that often must take place.

Returning again to the ESAT example, the relative merits of each of the four concepts identified above are considered. Figure 5.41[58] describes the four ADACS architectures with a brief critique of each. The key requirement here is the pointing accuracy, customer input requirement 2.2.8 from Table 5.1. That requirement states that the spacecraft bus shall point the telescope to an accuracy of ±0.01° on all three axes. The only architecture that has the capability of meeting such a requirement is the zero momentum design. Therefore, the zero-momentum concept is the design of choice.[59]

IV. Optimization and Tailorability

A. Optimization

Similar to the argument just asserted for Trade Analyses, optimization techniques are myriad and can be very specialized. It is not the purpose of this book to discuss various optimization approaches. Rather, the purpose is to describe where optimization occurs in the process and how it impacts the overall development process.

[58] Attitude control accuracy numbers are taken from Larson, Wiley J. and James R. Wertz, *Space Mission Analysis and Design*, Kluwer Academic Publishers, Dordrecht, The Netherlands, 1992, Table 11-4, p. 346.

[59] In this simple example the selection was made purely on the basis of technical performance. As has been discussed above, in the real world cost and schedule must also be considered. Note also that, as is often the case, there are several ways to implement the selected system. A zero momentum attitude control system is no exception and this will likely lead to more trade-offs.

Architecture

	Gravity Gradient	Spin Stabilization	Bias Momentum	Zero Momentum
	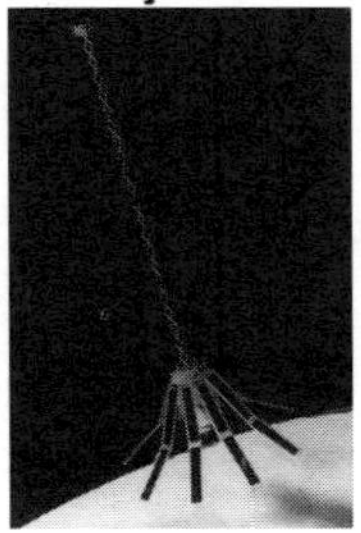 GEOSAT *Photo Courtesy APL JHC*	 DSP Spacecraft *Photo Courtesy TRW*	 TDRS *Photo Courtesy TRW*	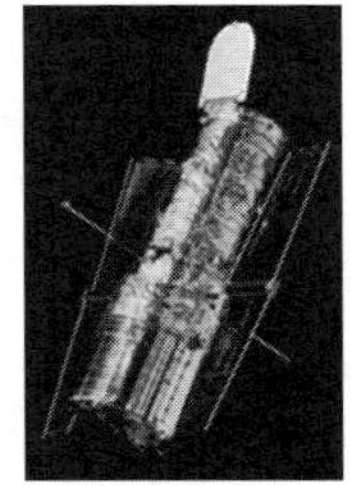 Hubble Space Telescope *Photo Courtesy NASA*
Design Elements	Passive Mass Dist	Aspect ratio Mass Balance Earth Sensor IMU, OBC Thrusters Mag Torqures	Earth Sensor Sun Sensor IMU, OBC Thrusters Momentum Wheel Mag Torquers	Earth Sensor Sun Sensor Star Tracker IMU, OBC Thrusters Reaction Wheels Mag Torquers
Accuracy	± 5° two axes	± 0.1° to ± 1°	± 0.1° to ± 1°	±0.001° to ± 1°
Assessment	Cannot meet 0.01 accuracy rqmt	Spin not OK Cannot meet 0.01 accuracy rqmt	Cannot meet 0.01 accuracy rqmt	Meets all rqmts

Figure 5.41 ADACS Candidate Architectures.[60]

[60] Note that while the design elements identified in the figure are representative of the design of the spacecraft shown, they do not necessarily represent the exact design of the actual spacecraft depicted in the figure.

Optimization, by definition, implies a change to the system. This change may be reflected in the requirements, which, as has been emphasized above, are organically connected to the implementation. Therefore, if optimization is to be done, the SDF provides a feedback to the Design and/or Requirements Development activities (Figure 5.42). While optimization can occur within virtually every element of the process, it is explicitly addressed here because the options being considered are "complete" in the sense that they have been defined to the point that technical, cost, and schedule criteria have been developed. At the other points in the process, each option is in the process of being developed so optimization occurs as a natural part of the Requirements Development and Synthesis activities through the iterations that occur.

B. *Tailorability*

Tailorability is usually a topic of discussion when a generic system engineering process is to be applied to a specific development program. At this point, it is asserted that, to the level of decomposition provided above, the SDF is applied to each hierarchical level and for each development phase. While not all activities represent significant effort in every situation, generally all are performed to some level of fidelity. Therefore, tailoring is *not* achieved by changing the process. Rather, it is achieved by:

- Modulating the kinds and extent of documentation required
- Modulating the level of detail and the scope of the activities performed
- Prudent partitioning of the system hierarchy to effectively satisfy program needs.

V. The Integrated System Development Framework

Figure 5.42 represents the fully-integrated, second-level decomposition of the basic SDF building block. It provides a two-tiered view of the SDF, defining the flow-down and feedback paths both within the same level and between levels of the system hierarchy.

Initial inputs are fed into the "Identified Work to Do" bucket. The first steps of the process involve both Work Generation activities as well as Rework Discovery activities. Because it is desirable to discover any problems with the requirements as early as possible, the "Analyze Requirements" activity is initiated. The Work Generation activities are also initiated. Progress is periodically assessed for convergence. If the Requirements Development activity is failing to converge, it may be the result of a discrepancy in the input requirements. A feedback to the "input" is shown in the figure, indicating that the customer should be consulted in order to review the input for consistency, clarity, and completeness. Nonconvergence may also be the result of less-than-adequate quality and/or insufficient effort focused on discovering rework. As discussed in Chapter 4, these are the control mechanisms with the highest leverage for enabling convergence.

If the Requirements Development activity is converging, the output data is passed along to the Synthesis activity. A key decision must be made regarding the timing of the release of the output data to the Synthesis activity. Data released prematurely will result in more rework generated than data released in good condition, albeit later on the timeline. Once data is passed to the Synthesis activity, the Work Generation activities commence. A decision block in the process asks if supporting work at the next level down is necessary. If the answer is "yes," requirements are generated and released to the lower-level activities. If the answer is "no," the effort moves to the "Integrate and Plan Verification" activity, where input from lower-level activities is integrated into the system design. As the design development ensues, the Verify Design activities commence. Most of the rework discovered will likely be redone in the Synthesis activity. However, there is the potential that some of the rework discovered may be Requirements Development rework. There is, therefore, feedback to the Requirements Development activity via the RD Rework decision box. It is shown as the "Discovered Rework" box. Note also the inclusion of a box called "Forced Rework." This occurs in those cases where the output from the preceding activity is valid in and of itself, but cannot be implemented or is difficult to implement for some reason. This situation can arise, for example, when technologies included in the design are obsolete or are otherwise unobtainable. Or, due to technical, cost, and/or schedule concerns, a change in the requirements set or current design is called for. Both the "Forced Rework" and "Discovered Rework" feedback boxes also occur as feedback from the lower levels to the upper-level Synthesis activity as shown in Figure 5.42. If the Synthesis activity is not converging, the same controls of quality and rework discovery effort can be adjusted to facilitate convergence.

As discussed earlier in this chapter, the "Do Trades" activity commences as needed after two or more designs emerge from the Synthesis activity. If optimization is necessary, there is feedback to the Synthesis activity.

Figure 5.43 depicts the basic SDF building block along its one-level-down decomposition. It is provided in order to emphasize the consistency of the decompositions, as described in the foregoing.

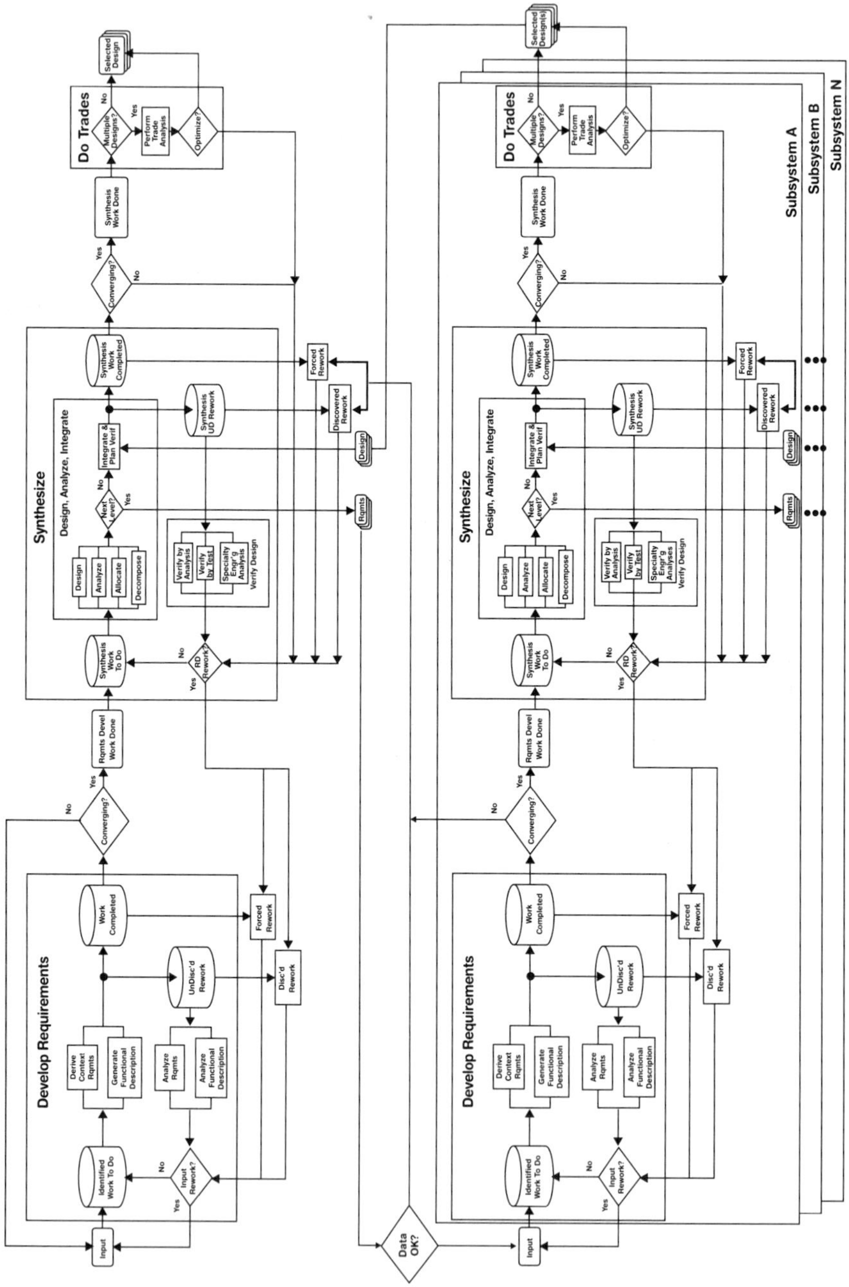

Figure 5.42 The System Development Framework (SDF), Second Level Decomposition.

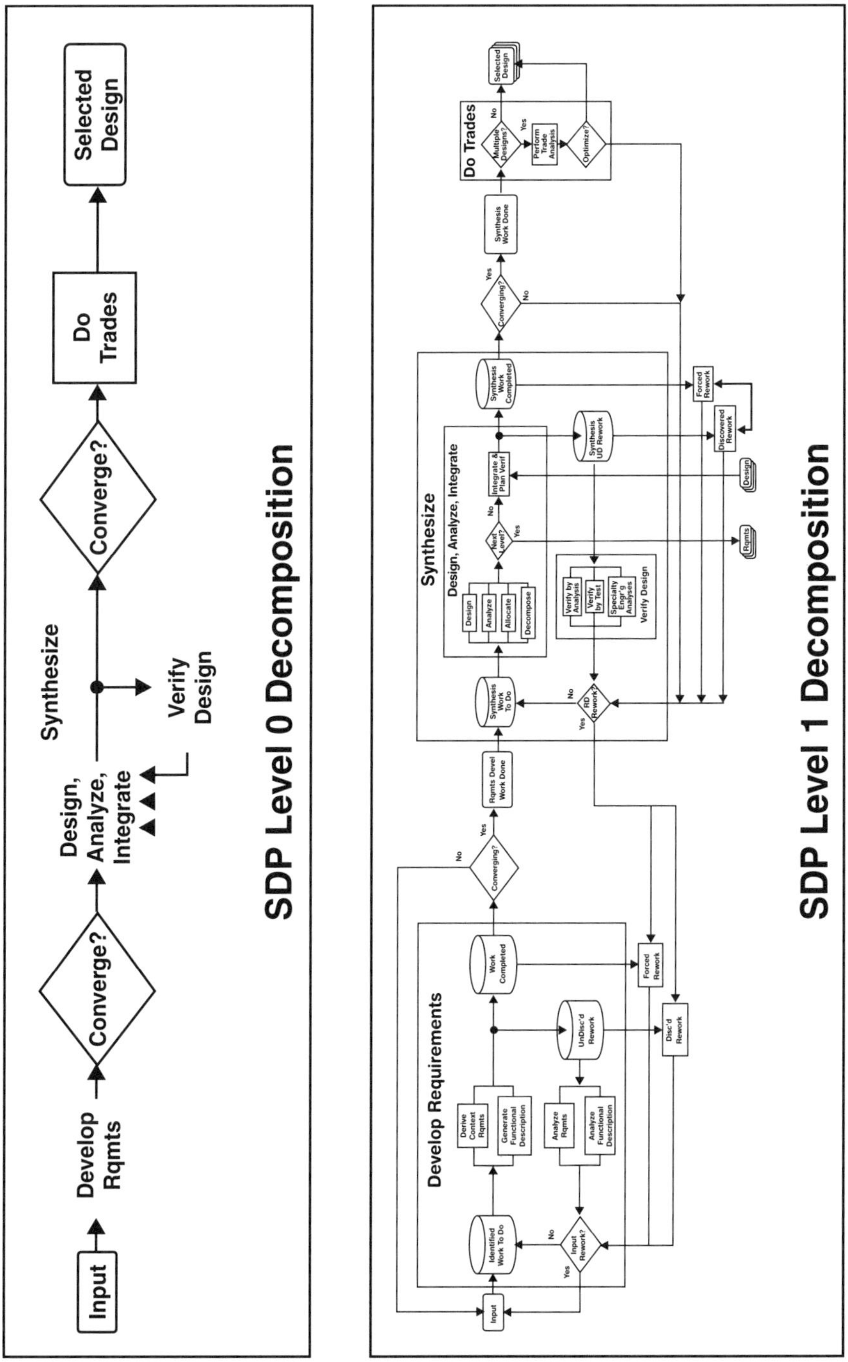

Figure 5.43 SDF Decomposition Consistency.

chapter 6

The System Development Framework — Managerial

It's easy to play any musical instrument:
all you have to do is touch the right key at the right time
and the instrument will play itself.
—Johann Sebastian Bach

The other teams could make trouble for us if they win.
—Yogi Berra

I. Integrating Technical and Managerial Activities

As Figure 6.1 illustrates, activities composing the management effort include Risk Management, development and tracking of metrics and technical performance measures, Configuration Management, etc. The technical effort comprises Requirements Development, Synthesis, and Trade Analyses. Because these activities are closely related, they must be coupled for efficient management of the development. The SDF can be used to effectively link these together.

II. Developing the Program Structure

How should the program be partitioned to enhance the efficiency of the development? How should information flow from one development team to another? Where does each development team get its requirements? Who should be responsible for identifying, characterizing, and controlling interfaces? The SDF can be used to address these key questions.

Logically, the first task in developing the program structure is to partition the system into its constituent elements. In the case of a precedented system, the first level decomposition is usually well understood. A low earth orbiting spacecraft bus, for example, is often partitioned into the following subsystems: structures and mechanisms, thermal control, electrical power,

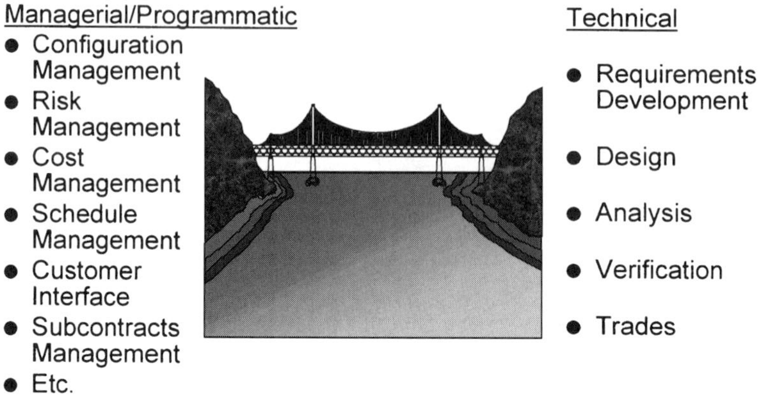

Figure 6.1 Technical and Managerial Activities[61]

attitude determination and control, propulsion, communications, and command and data handling. It is often the case that the development teams are organized in the same way. A significant benefit of doing this is that the interfaces and roles and responsibilities are well understood. In the case of an unprecedented system, the partitioning of the functional teams should be guided by the current system description to provide a first-cut hierarchical arrangement of functional teams.

With regard to information flow, it is often the case that problems within a system occur at an interface.

Key Point

> Therefore, a key objective in partitioning the system into subelements is the minimization of the number of interfaces between partitioned elements.

The Design Structure Matrix (DSM) can be used to determine the best partitioning of the program elements as well as the functional teams. Figure 6.2 illustrates a typical DSM. Each system element is represented in the matrix down the first column and across the first row. The interfaces between elements are indicated by an "X" where the appropriate column and row intersect. The diagonal is obviously left blank. Partitioning is indicated by a box drawn around the group of elements that make up each

[61] Adapted from Adamsen's presentation charts presented at the 1996 INCOSE Symposium. Cf. Rochecouste, who develops a similar listing of activities. Systems Engineering Process/Activities: Requirements Analysis, Functional Analysis, Design Synthesis and Specifications, Subsystem Integration, System Test and Evaluation, etc. Technical Management Activities: Technical Planning, Requirements Management, Configuration Management, Design Reviewing, Technical Risk Management, Technical Performance Measurement, etc. Rochecouste, Hervé. *A Systems Engineering Capability in the Global Market Place*. P004.

	A	B	C	D	E	F	G	H	I	J	K	L	M	N	O
A	•														
B		•	X						X						
C			•			X			X		X				
D			X	•	X							X			
E			X	X	•										
F		X	X	X		•			X						
G				X	X	X	•	X							
H			X			X	X	•	X		X				
I					X	X	X	X	•	X				X	
J					X	X	X	X		•					
K		X				X		X	X		•	X			
L			X			X		X			X	•	X		
M			X			X						X	•	X	
N														•	X
O															•

Figure 6.2 The Design Structure Matrix (DSM).

subsystem. A goal is to arrange the order of the elements such that the interfaces between them are minimal in number.[62]

Obviously in the case of precedented systems, there may be limited flexibility in terms of repartitioning established subsystems. However, whenever possible, interfaces between elements should be minimized. This applies to the creation of development teams on the program. If the system design has been efficiently partitioned so as to minimize interfaces, this will likely provide an efficient model for partitioning of functional teams as well. For this reason, it is advocated that the team structure should follow the structure of the partitioned system insofar as it makes sense to do so.

Now that the question regarding how to partition the program hierarchy has been addressed, the question concerning how these partitioned elements ought to interact is discussed. It is suggested that the logical flow of information between the partitioned elements should be provided by the control logic defined in the SDF.

Figure 6.3 illustrates the development of the program structure from the existing system design partitioned so as to minimize the interfaces between system elements. It also uses the "control logic" of the SDF to define the information flow paths within the total system hierarchy.

[62] A thorough treatment of the subject of Design Structure Matrix methodology is beyond the scope of this book. For more information, see Steven D. Eppinger, Daniel E. Whitney, Robert P. Smith, and David A. Gebala, A Model-Based Method for Organizing Tasks in Product Development, *Research in Engineering Design*, 6:1-13, 1994, and Kent R. McCord and Steven D. Eppinger, Managing the Integration Problem in Concurrent Engineering, MIT Sloan School of Management, Working Paper Number 3594, August 1993. See also Rosaline K. Gulati and Steven D. Eppinger, The Coupling of Product Architecture and Organizational Structure Decisions, MIT Sloan School of Management, International Center for Research on The Management of Technology, Working Paper Number 151-96.

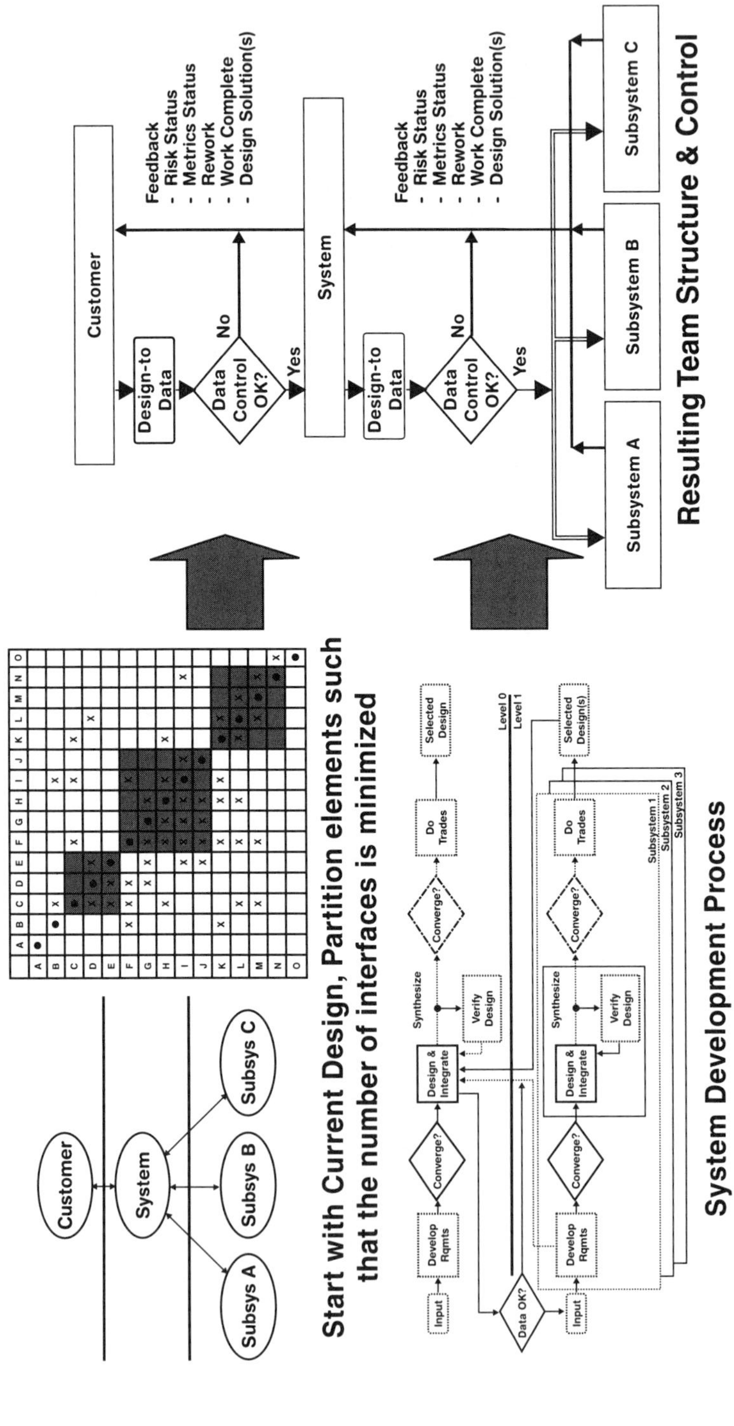

Figure 6.3 Program Structure Development.

There are several reasons for applying the control logic provided by the SDF to the program structure:[63]

- Defines activity flow and information flow between and within development teams
- Provides basis for requirements, design, and decision database structure
- Identifies interfaces between tiers of system hierarchy
- Defines team roles and responsibilities in context of total system hierarchy
- Defines activities more precisely enabling more precise progress measurement

III. Interaction in the Logical Domain

Figure 6.4 illustrates the horizontal and vertical interfaces between the various functional teams in the program hierarchy. These interfaces are derived directly from the SDF. All "design-to" data is passed to a lower level from the team at the level directly above. This is also true of feedback. All such data is communicated to the level directly above. Data travels horizontally through the team at the level above. It is the responsibility of that team to disseminate information to the appropriate teams under its jurisdiction. Remember, it is formal interfaces that are being addressed here. Of course, it is desirable for teams on the same horizontal level to communicate and interact.

Key Point

> All design data, however, must pass through and be coordinated by the team at the level above. This is important in order to maintain control of the information and to ensure that all information is communicated accurately and in a timely manner.

It is also important to maintain such control when changes occur. The appropriate reviewers must be consulted so that all impacts resulting from the change are properly assessed. Any modifications to the design made to accommodate such changes must be coordinated so as not to introduce new problems into the design.

Each lower level element functions semi-autonomously from its next level up element as long as it stays within its allocations (i.e., prescribed boundary conditions). Periodic status is provided to the level above in terms of technical performance measures, risk assessments and mitigation approaches, design issues and solutions, etc. If an allocation is violated, the

[63] Adamsen (1996), pp. 1093-1100.

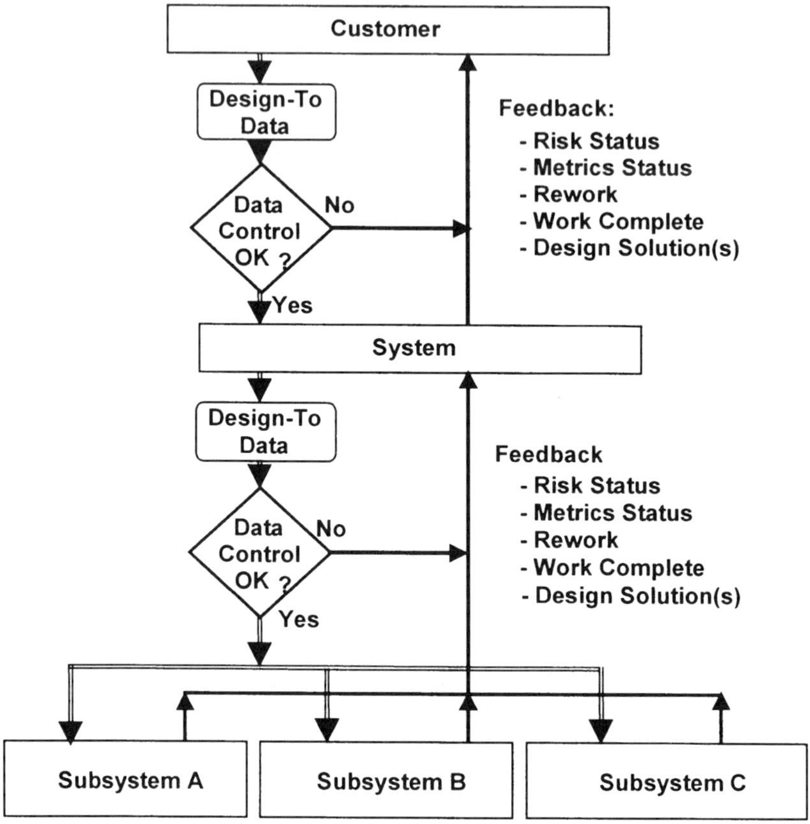

Figure 6.4 Program Team Interactions.[64]

next level up element becomes involved to either reallocate or modify the design at its level.

IV. Interaction in the Time Domain

Having described how the SDF drives the interactions of teams in the logical domain, the question of how it functions in the time domain is now addressed. Figure 6.5 depicts the SDF in the time domain. Along the timeline, the number of options converges to a single solution while the level of detail definition increases. As discussed in previous chapters, the time domain view describes how the generation of data evolves as a function of time. This includes prescribing the level of fidelity of each output required at a particular point on the program timeline for each tier of the hierarchy. As discussed previously in Chapter 3, in order to proceed along the program timeline with minimal risk, it is necessary to "incrementally solidify" key requirements by program milestone.

[64] *Ibid.*

Hierarchical Tier	Award	First	Second	Third	Fourth
Level 0 Rqmts Level 0 Design	>70% >50%	>90% >70%	Update >90%	Update Update	Update Update
Level 1 Rqmts Level 1 Design	>50% --	>70% >50%	>90% >70%	Update >90%	Update Update
Level 2 Rqmts Level 2 Design	-- --	>50% --	>70% >50%	>90% >70%	Update >90%
Level 3 Rqmts Level 3 Design	-- --	-- --	>50% --	>70% >50%	>90% >70%

Figure 6.5 Time Domain View.

Key Point

> It is highly desirable to determine which requirements are needed from the customer and when, in order to maintain progress with minimal risk. The goal is to include these critical need dates in the contract so that if there is delay, a cost and schedule scope change can be negotiated.

Figure 6.5 illustrates how requirements and design data are progressively generated. Four stages of increasing fidelity are illustrated: 50%, 70%, 90%, and update.[65] Requirements lead the development of the implementation by one stage of fidelity. First, the system level requirements are generated to a fidelity level of, say, 70%. Once the requirements set has reached this level (or a level of acceptable risk), the Synthesis activity can begin with an acceptable level of risk. As the Synthesis activity progresses toward an increasing level of fidelity, the requirements are also increasing in fidelity as they are validated against the developing design.

Also illustrated is the progression of the Development activity down the hierarchy as a function of time. As upper levels become increasingly stable, lower-level requirements and Synthesis activities are initiated at acceptable levels of risk. Thus, the key criterion for vertical (e.g., system to subsystem) and horizontal (e.g., movement from Requirements Development to Synthesis or passing a Major Milestone Review) progression is attaining to an acceptable level of risk. This suggests that the more accurately risk can be quantified, the better the program can be managed in terms of meeting cost and schedule goals.

[65] In this context, fidelity refers to the completeness of the requirements or the certainty of the requirements in terms of stability or risk. The quantification of the status is admittedly subjective and represents a relative measure and not an absolute one.

Key Point

> The above discussion indicates the necessity of solidifying the requirements to some level before design activities are initiated in order to minimize the risk of a nonconverging or slow-to-converge Synthesis effort. It further illustrates the necessity of solidifying the design before requirements for the next level down are passed along, thereby initiating the development activities at that next level.

This is one cause of overruns on development programs. Requirements are not stabilized prior to initialization of the synthesis activity. Or, the design is not stabilized prior to initializing the next-level-down development activities. This implies that a structured approach to system development is needed.

V. A Note on Complexity

Systems comprise various subsystems which also comprise several sub-subsystems and so on. It is intuitively apparent that complexity grows as more subsystems and tiers are added to the system. Furthermore, growth in complexity appears to be non-linear in that it grows so quickly on many development programs.

It might be helpful to quantify the growth of complexity in order to understand its impact on a program. Figure 6.6 depicts four tiers in the hierarchy, with three subsystems below each system above. It is clear that growth in complexity is an exponential function. It has been assumed that each subsystem comprises only three subsystems. This is quite conservative as many systems comprise seven subsystems or more. For example, a typical spacecraft can have seven or eight subsystems, while an aircraft can have more still.

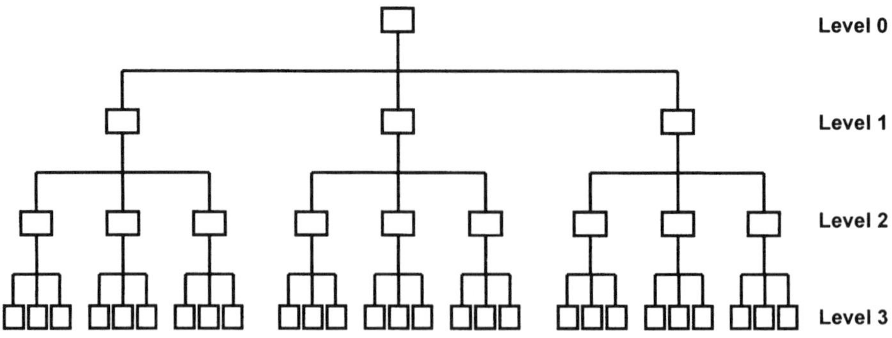

Figure 6.6 Exponential Growth in Complexity.

To put this in mathematical terms, if "S" represents the number of subsystems per system and "n" the nth tier in the hierarchy, then the following relationship emerges:

$$\textit{Total System Elements} = \sum_{n=0}^{m} S^n \qquad 9910(6.1)$$

The Total System Elements (TSE) refers to the total number of hardware and software pieces of the system. Therefore, as the TSE grows, the overall complexity also grows as does the energy needed to develop each element. As has been suggested previously, in this context energy refers to manpower and all other resources needed to perform the development of all the system elements.

Figure 6.7 is a graphical depiction of the above relationship. One curve is plotted for the assumed number of subsystems below each system. Figure 6.6 is a picture of the case where three subsystems are assumed for each system. The total number of elements included in Figure 6.6 is 40. This is represented in Figure 6.7 by the curve labeled "3" in the legend. It is apparent from the graph that complexity increases rapidly as subsystems per tier are added and as tiers are added to the system hierarchy.

VI. Major Milestone Reviews

A central motivation for conducting a technical review is to gain consensus with all stakeholders that the requirements and design, at the level of the review, are of sufficiently low risk or acceptable risk, so as to continue toward the next milestone. There are relatively few complex system development efforts that are accomplished on schedule and within the proposed cost. One reason may be that the purpose of each Major Milestone Review is not clearly defined. Both the customer and the contractor must know when critical requirements are needed in order for the development activity to proceed with an acceptable amount of risk. As discussed previously, if requirements or implementation are unstable at upper levels, the risk induced can be significant. Therefore, the following is offered as suggested objectives for each review.[66]

First Major Milestone Review — This review is conducted to solidify the system requirements in the form of a configuration-controlled system-level specification and the Interface Control Document (ICD) that defines *external* interfaces before significant resources are invested in developing the system-level design. This review focuses on understanding the total context in which the system must function over its full life cycle. It assesses the risk profile of the program in order to determine whether or not to proceed toward the Second Major Milestone Review.

[66] See Appendix C for SDF-derived major milestone review criteria.

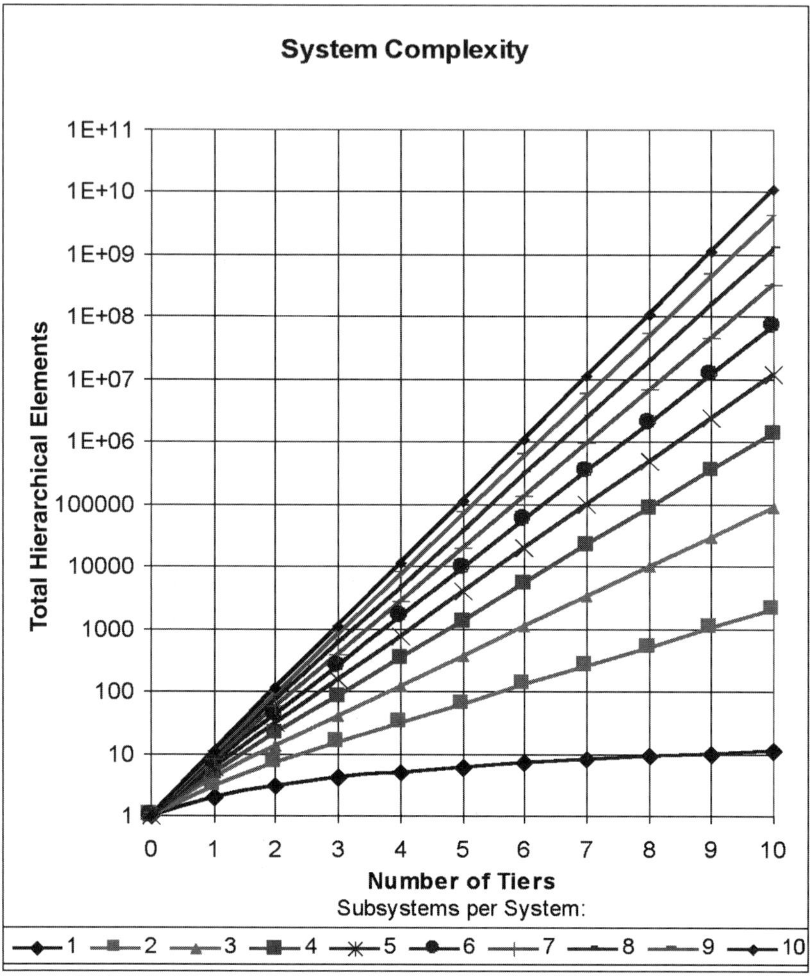

Figure 6.7 Complexity Growth.

Second Major Milestone Review — This review is conducted to establish the baseline system-level architecture by configuration controlling the system block diagram, ICDs that characterize and control system-level *internal* interfaces, and the Operations Concept. Successful completion of this review results in the release of the configuration-controlled specifications for the next level system elements. It assesses the risk profile of the program in order to determine whether or not to proceed toward the Third Major Milestone Review.

Third Major Milestone Review — This review is conducted to establish the subsystem baseline designs by configuration controlling the subsystem block diagrams and ICDs that characterize and control subsystem-level internal interfaces. Successful completion of this review results in the release of the configuration-controlled specifications for the next level system elements. It assesses the risk profile of the program in order to determine whether or not to proceed toward the Fourth Major Milestone Review.

Fourth Major Milestone Review — In general, this review is conducted to establish the component-level baseline designs and any necessary next-level-down specifications by configuration controlling the component block diagrams and any necessary ICDs that characterize and control component-level internal interfaces. Successful completion of this review results in the release of any remaining and necessary configuration-controlled specifications for the next level system elements. It assesses the risk profile of the program in order to determine whether or not to proceed toward the build phase of development. Note that in a very large system with many hierarchical tiers it may be necessary to add more major milestone reviews. In such a case, the basic flow and content of the reviews remains the same as indicated in the preceding.

VII. What About Metrics?

Certainly, a central concern of any program manager is how to determine the true status of the development. Is the right amount of progress being made? What is the projected cost to complete the development of the system? When will the development phase be completed? What are the risks and their potential impacts?

Meaningful metrics in the area of system engineering have been difficult to define. One reason for this difficulty may be the lack of a well-defined system engineering process that can be consistently applied across a broad range of programs. It is virtually impossible to implement a process that is inadequately defined and it is likewise difficult to measure progress against such a process.

The SDF is defined in sufficient detail to enable the implementation of meaningful metrics. Each major activity (i.e., Requirements Development, Synthesis, and Trades) and each subactivity (e.g., requirements analysis, functional analysis, design, allocation, analysis, integration, verification, etc.) is allocated cost, schedule, manpower, computer, and other resources. Actual consumption of these resources is tracked against the allocated resource plan to measure progress. These data are cataloged in a database. As more programs are performed, the metrics are refined in terms of the necessary allocation of resources for each activity. The database becomes the benchmark against which future programs are measured. In order to ensure that apples are compared to apples and not to oranges, this approach necessitates that each program implement the same SDF. Such a database would be useful for estimating costs for new proposals and for internal estimates of completion.

chapter 7

A Potpourri of SDF-Derived Principles

Whether we will philosophize or we won't philosophize, we must philosophize.
—Aristotle

If I wished to punish a province, I would have it governed by philosophers.
—Frederick II, the Great

The following principles have been derived during the course of study that lead to the development of the SDF described in the preceding chapters. These principles or heuristics have been collected under several broad categories, none of which are absolute.

I. General

1.1 A system architect is responsible to define the interfaces internal to his/her system, but he/she may not have control over those interfaces external to his/her system.

1.2 A sound architecture and a successfully managed program consider the total context within which the system must function over its full life cycle. It is not sufficient to develop an architecture for a deployed system that does not consider all contexts in which the system must function: manufacturing, integration and test, deployment, initialization, normal operations, maintenance operations, special modes and states, and disposal activities.

II. Risk

2.1 Acceptable risk is a key criterion in deciding to move from one activity to the next. The challenge is to quantify risk in an accurate and meaningful way.

2.2 Allocation of resources is a key basis by which to measure and monitor risk. Risk in a system development activity can result from insufficient margin in technical, cost, and/or schedule allocations.
2.3 Program risk increases exponentially with requirements instability. Because upper-level requirements drive lower-level designs and requirements, and because the number of system elements increases exponentially down the program hierarchy, a few unstable top-level requirements can affect many lower-level system elements.
2.4 Risk is difficult to manage if it has not been identified.
2.5 In conceptual architecting, the level of detail needed is defined by the confidence level desired or the acceptable risk level that the concept is feasible.

III. Functional Analysis

3.1 A function cannot be decomposed without some reference to implementation. That which enables the decomposition of a function is knowledge and/or assumptions about its implementation.
3.2 A functional decomposition must be unique. Prescribed functionality describes "what" the system must do and, in order to be self-consistent, that functional description must be unique.
3.3 Function partitioning is not unique. There are many ways to partition functions.
3.4 The functional definition must include both functionality and associated performance in order to be useful in implementation. It is necessary but not sufficient for implementation to define only required functionality. Performance must also be prescribed in order for a function to be implemented in a meaningful way.
3.5 Within the same tier, the "what" (Requirements Development) drives the "how" (Synthesis activity). Form must follow function. Implementation, by definition, performs a function. It is not rational to try to determine "how" to perform a function that has not been identified.
3.6 With respect to interaction between hierarchical tiers, the "how" above drives the "what" below. This is a very important principle and has implications in areas such as specification development. When a customer or next-level-up design team defines functionality in a specification several tiers down, the probability of introducing problems increases because the intermediate decompositions may not be consistent with the prescribed requirements.

IV. Allocation

4.1 Margin unknown is margin lost. In order to optimally manage a system development, all system margin must be known to the architect having authority to allocate it. It is generally cost-effective to reallocate resources to handle issues. In order to do this effectively, the architect needs to know where the margin resides.

4.2 During an architecting effort, cost and schedule should be allocated and managed just as any other technical resource.

V. Process

5.1 Process understanding is no substitute for technical understanding. This is exemplified by the principle that a function cannot be decomposed apart from implementation. It is the technical understanding of the architecture implementation that facilitates the decomposition.

5.2 Before a process can be improved it must be described.

5.3 Tools should *support* the system development process, not *drive* it.

5.4 A central purpose of the SDF is to provide every stakeholder with a pathway for understanding the system and the state of its development.

5.5 There is a necessary order in which technical activities must be performed. Some notion of "what" the system must do must precede any effort to determine "how" to do it. Some notion of "how" the system will be implemented must precede any determination of "how well" it performs. Because trade analyses are based upon cost, schedule, and/or technical criteria, they must follow the Synthesis activity. A trade analysis cannot be performed without some definition of implementation. Trade analyses are based upon cost, schedule, and technical criteria. These cannot be determined without some relation to implementation.

5.6 Any technical activity can be categorized as either a "what," "how," "how well," "verify," or "select" activity. This is the primary organizing principle of the generalized SDF.

5.7 Describing the System Engineering Process in both the time domain (output evolution) and the logical domain (energy distribution) facilitates its application to many contents.

5.8 A system is "any entity within prescribed boundaries that performs work on an input in order to generate an output."

5.9 Given the above definition of "system," the same system development process can be applied at any level of design development.

5.10 The development process must include not only the deployed system but also all necessary supporting entities. A sound architecture involves the architecture of supporting system elements as well as the element that actually performs the mission functions directly.

5.11 Interface control is always the responsibility of the element in which the functional decomposition occurs. Interface identification and characterization must involve the elements at which they occur, but their control is the responsibility of the element above which the interface occurs.

5.12 The technical process of architecting a system should drive the approach for managing the system development. Interfaces can (and should be) defined by the technical process. These help determine roles and responsibilities of teams, information flow and control, subcontract boundaries, etc.

VI. Iteration

6.1 Iteration generally occurs for one of three reasons: optimization, derivation, or correction. Optimization of a design (at some level of fidelity), by definition, results in a change to that design. Where change is necessitated, feedback and/or iteration occurs. This, of course, is distinct from optimization techniques that are used to develop a design based upon known parameters (e.g., linear, non-linear, and integer programming techniques). Therefore, when optimization is necessitated there is often feedback to the Requirements Development and/or Synthesis activities. Derivation refers to those situations where lower-level design must be done in order to provide the needed confidence level of the architecture at the level above. Correction covers that broad category of issues that arise as a result of errors.

6.2 Always try re-allocation before redesign. It is usually less expensive to reallocate resources than it is to redesign.

6.3 The cost of rework increases exponentially with time. It is relatively easy to modify a specification. It is more difficult to modify a design. It is still more difficult to modify several levels of design and decomposition. It is yet more difficult to modify hardware when it is in manufacturing, still harder during integration and test, still harder once deployed, and so on.

VII. Reviews

7.1 A central purpose of a Major Milestone Review is to stabilize the design at the commensurate level of the system hierarchy. At the first major milestone review, for example, the primary objective ought to be to stabilize the system level architecture and the lower-level requirements derived from it. This facilitates proceeding to the next level of design with an acceptable level of risk.

VIII. Metrics

8.1 A metric's "coefficient of elusivity" is proportional to the definition resolution of the process it is supposed to measure. The more accurately a process is defined and implemented, the more easily it can be accurately measured. This is a significant contributor to the difficulty of defining useful system engineering metrics. Lack of a sufficiently detailed and universal process makes universally applicable metrics difficult to define.

IX. Twenty "Cs" to Consider

9.1 Control — Who's minding the store and how?
9.2 Context — How does this system fit into the next larger system?
9.3 Commonality — Can it be the same for all?
9.4 Consensus — Does the customer agree with our interpretation?
9.5 Creativity — Have all the creative solutions been considered?
9.6 Compromise — Are all system parameters properly balanced?
9.7 Change Control — Are all system impacts understood?
9.8 Configuration Management — Is everyone working to the current configuration?
9.9 Comprehension — Is the system understood in terms of what it must do and how it works?
9.10 Characterization — Has the required system functionality, performance, and rationale been defined and communicated?
9.11 Coherence — Does the system function as an integrated whole?
9.12 Consistency — Have all conflicts been resolved?
9.13 Completeness — Has the system been fully defined?
9.14 Clarity — Have all ambiguities been removed?
9.15 Communication — Is there good communication between all stakeholders?
9.16 Continuity — Will successors know why it was done this way?
9.17 Cost Effectiveness — Is the system over-designed?
9.18 Competitiveness — Can it be done better, faster, and/or cheaper?
9.19 Compliance — Does the system do what it is required to do?
9.20 Conscience — Have we done our best?

X. Suggestions for Implementation In Industry

Industry has been plagued with the "process *du jour* syndrome" for a number of years. As a result, there is significant reluctance on the part of many engineers to support yet another three-lettered process to fix all program problems. These attitudes often have merit. Many of these three-lettered processes are not worth the paper on which they are written.

A different tack to implementing this structured approach to complex system development in industry is suggested. First, implementation of the SDF is "tailored" to the specific application by identifying up front what the required inputs and outputs will be for each SDF activity. This can be accomplished with the worksheet, provided in Appendix A, which provides a framework for identifying outputs as well as release status and risk assessment. Second, Exit Criteria is developed for each Major Milestone Review. These criteria are derived directly from the outputs identified in Chapter 5. The required fidelity or "completeness" of each output is defined as a function of time along the program timeline. In addition, the purpose for each major review is clearly defined, as discussed in Chapter 6. Appendix C provides an example of Exit Criteria for a typical Major Milestone Review derived from the outputs of the SDF.

For each Major Milestone Review, the program must produce the generalized output defined in Chapter 5. In so doing, the structured approach is followed by default. There are several reasons for moving the implementation strategy in this direction. While SDF training courses have been, in general, very well received by engineers in industry, there is still a significant element of the "Don't tell me how to do my job" attitude. This is despite the fact that even a cursory evaluation of Chapter 5 must acknowledge that it is only a framework that is constructed. A second reason for this approach is that the outputs required are generally necessary. Most would agree that the outputs identified are a necessary element of most any development program. How the outputs are generated is not prescribed, nor is the format in which they must be presented dictated. It is simply required that they be completed and presented at the Major Milestone Reviews.

appendix A

Small Product Development and the SDF

Ulrich and Eppinger define a Product Development Process (PDP) that focuses on small product development. There are two significant differences between the PDP and the SDF. First, for simplicity, the PDP avoids a multitiered hierarchy. Second, the rework cycle is not explicitly integrated into the PDP. These are some of the key differences between small product development and complex system development. It is not necessary to encumber a small-scale development task with a rigorous hierarchy of subsystems and components since it would take more energy to manage those tasks than to perform them. Also, since rework does not ripple through a complex hierarchy, its adverse impacts are not as great.

I. Mapping in the Logical Domain

Ulrich and Eppinger's Front-End Development Process is shown in Figure A1. As Figure A2 indicates below, its key activities map directly from the Logical Domain view of the SDF in the categories of Requirements Development, Synthesis, and Trades. The "Refine Specifications" activity is handled via feedback loops in the SDF. The "Plan Remaining Development Projects" activity would be considered a management activity in the SDF; elements of this task would also be covered in the Synthesis activity.

Logical Domain mapping of the PDP to the SDF is straightforward. The organizing concept, discussed in Chapter 3 above, readily applies to the PDP.

II. Mapping In the Time Domain

Figure A3 depicts Ulrich and Eppinger's Product Development Process. It maps directly to the Time Domain view of the SDF, as shown in Figure A4. Phases 1 to 4 map directly to the full life cycle view of the SDF illustrated earlier in Chapter 3.

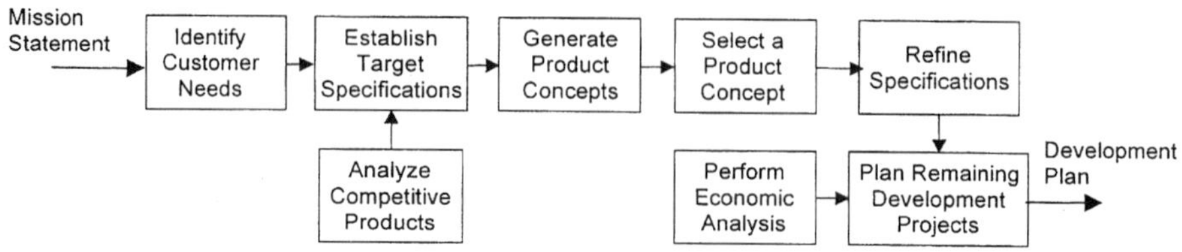

Figure A1 Ulrich and Eppinger's Front-End Process.[67] (Courtesy McGraw Hill, used with permission.)

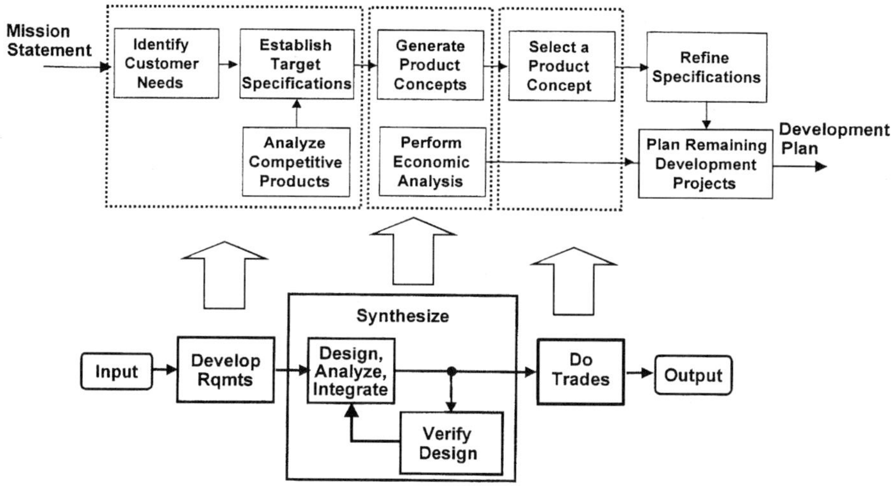

Figure A2 Mapping PDP to SDF in Logical Domain.

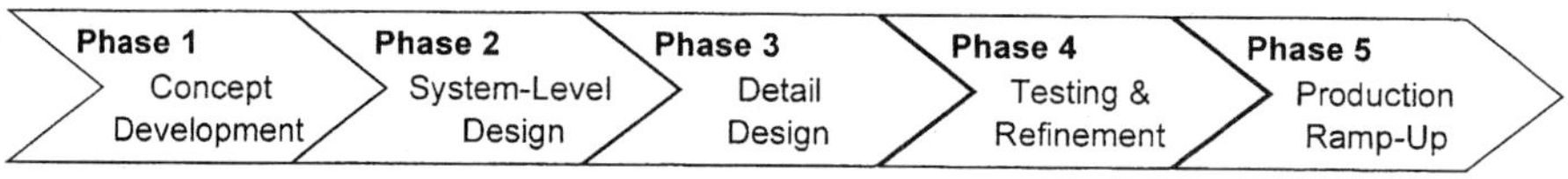

Figure A3 Ulrich and Eppinger's Product Development Process (PDP).[68] (Courtesy McGraw Hill, used with permission.)

Ulrich and Eppinger's small-scale development process represents a solid distillation of the SDF that focuses on the needs of simple systems. Multiple tiers are avoided and the rework cycle is not explicitly included. This brief discussion illustrates the utility of defining the Design Development Process in terms of both the Time and Logical Domains in order to preserve universality of application. This general rule applies to both simple and complex design development contexts.

[67] Adapted from Ulrich, Karl T. and Steven D. Eppinger, *Product Design and Development*, New York: McGraw-Hill, 1995, p. 18.
[68] Adapted from Ulrich and Eppinger (1995), p. 9.

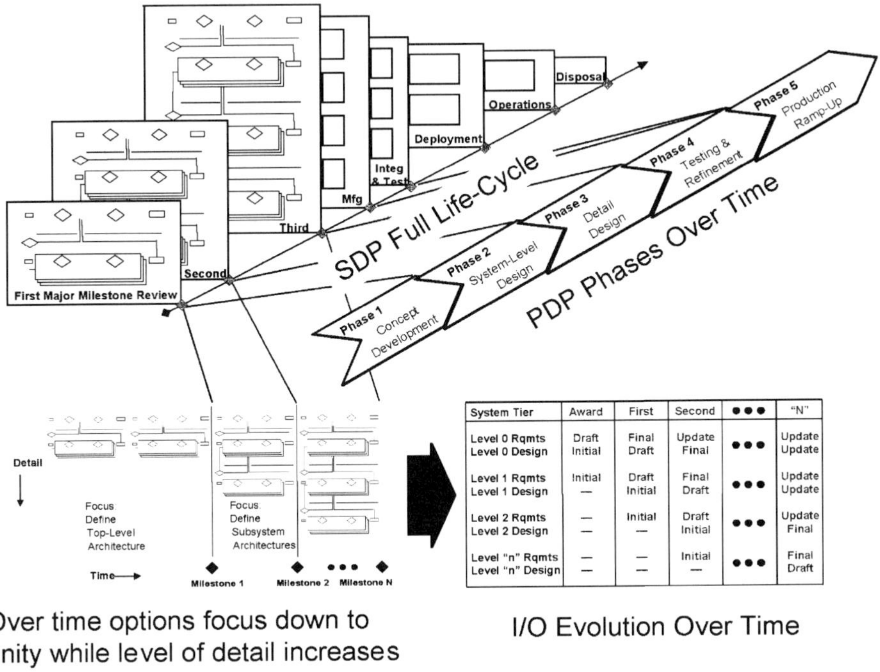

System Tier	Award	First	Second	●●●	"N"
Level 0 Rqmts Level 0 Design	Draft Initial	Final Draft	Update Final	●●●	Update Update
Level 1 Rqmts Level 1 Design	Initial —	Draft Initial	Final Draft	●●●	Update Update
Level 2 Rqmts Level 2 Design	— —	Initial —	Draft Initial	●●●	Update Final
Level "n" Rqmts Level "n" Design	— —	— —	Initial —	●●●	Final Draft

Figure A4 Mapping PDP to SDF in Time Domain.

appendix B

Tailored Documentation Worksheet

As mentioned previously, the SDF is "tailored" by defining the necessary outputs that will be generated during the development. It has also been mentioned that the output of one activity becomes the input to the subsequent activity. The ability to produce high-quality, low-risk output is directly dependent upon the quality of the input. Unstable or ill-defined inputs generally result in higher risk outputs. Figure B1 illustrates a Tailored Documentation Worksheet (TDW). It provides a mechanism for identifying specific documents that will be generated and for assessing the risk involved by using the data received. This same data can be used to develop program costs.

A risk assessment should be performed by the entity receiving the input to quantify the risk associated with proceeding. A decision should then be made by the entity having jurisdiction over the element in question as to whether or not the risk is acceptable. The objective of this approach is to identify system risk in order to properly manage it.

Major Milestone Review: 1_____, 2_____, 3_____, 4_____					
Activity	Output			Decision	
	Document Title	Document State (Draft, Preliminary, Final)	Risk Level (Low, Meduim, High)	Approved/ Rejected	Date
Requirements Development					
Derive Context Requirements					
Generate Functional Description					
Analyze Requirements					
Analyze Functional Description					
Synthesis					
Design					
Allocation					
Analysis					
Functional Decomposition					
Inter-Level Interface					
Integration					
Development Testing					
Test Planning					
Producibility Analysis					
Testability Analysis					
Other Specialty Engineering					
Trade Analyses					

Figure B1 Tailored Documentation Worksheet.

appendix C

SDF-Derived Major Milestone Review Criteria

This appendix provides a strawman outline for the contents of the first three Major Milestone Design Reviews. Not all the identified activities and outputs are necessarily pertinent to every development program or each tier of development on any particular program. Therefore, this appendix is intended only as a guide and must be tailored to meet the needs of any specific program effectively.

The reader will notice that nearly the same activities and output data are identified for each of the three reviews. This is in keeping with the application of the SDF to each element of the system hierarchy.

I. First Major Design Review

A. System-Level Requirements Development

1. Customer-to-Contractor Specification Correlation Matrix — Confirm Traceability
   a. Identification of all TBDs, TBSs, and TBRs and assessment of criticality
   b. Identification of requirements ambiguities
2. Identification of all other requirements and constraints impinging on the system
   a. Technical
   b. Cost
   c. Schedule
3. Verification Matrix identifying verification method and location in system build-up
4. Definition of System Context
   a. Definition of all program phases
   b. Identification and characterization of all external interfaces by program phase
      1) High rate data

        2) Low rate data
        3) RF signals
        4) Test and diagnostic interfaces
        5) Timing, sync signals
        6) Primary and redundant power
        7) Mechanical and thermal interfaces
        8) Etc.
    c. Definition of all environments by phase
    d. Identification of all critical events by phase
    e. Identification of all modes and states by phase
    f. Operations Concept
        1) Logistics plan
        2) Maintenance plan
        3) Operability plan
        4) Etc.
    g. Mission timeline
5. System Specification Tree
6. System Plan Tree
7. Functional Definition of the System Architecture (used to correlate system-required functionality and performance with new and heritage designs implemented)
    a. Functional block diagram for each program phase
        1) Functions required with associated performance traced from top-level system specification
        2) Inputs
        3) Outputs
        4) Noise sources
        5) Control functions identified

B. *Synthesis*

Design

1. Parametric Analyses Leading to Baseline Architecture
2. System Configuration Description
    a. System block diagram — Electrical architecture
    b. System layout/drawing(s) — Mechanical architecture
    c. System drawing tree
    d. System ICD tree
3. Identification of New Technologies Implemented
    a. Risk assessment and mitigation approaches
    b. Tailoring required for system application
4. Identification of Heritage Elements Implemented
    a. Internal interfaces identified and characterized by phase
    b. Changes required to meet system application

5. System Budgets
   a. Technical: mass, power, baseband, RF, memory, throughput, etc.
   b. Cost and schedule
   c. Present margin, contingency, and reserve and location in the system
6. Subsystem Specifications
   a. Traced from system specification
   b. Listing of TBDs, TBSs, TBRs
   c. Identification of all other known requirements issues
   d. Allocation of functionality, performance, cost, schedule
7. Identification and Characterization of Internal Interfaces
   a. High rate data
   b. Low rate data
   c. RF signals
   d. Test and diagnostic interfaces
   e. Timing, sync signals
   f Primary and redundant power
   g. Mechanical and thermal interfaces
   h. Etc.
8. System Risk Analysis
   a. Identification — Internally and externally driven
   b. Assessment of likelihood and potential impacts to system
   c. Mitigation approach(es)
9. Configuration Control System
   a. Configuration control board
   b. Action item status
   c. Change notice status
   d. System database status: wire lists, budgets, etc.
   e. Interface control
      1) Internal
      2) External
      3) ICD status
10. Operations Concept
11. System Optimization
    a. Sensitivity analyses — How sensitive is the system to potential changes?
    b. Requirements impact assessment

System Verification

1. System Simulation
   a. Correlation to specification
   b. Margin
   c. Etc.
2. FMECA (Failure Modes Effects Criticality Analysis)
3. FDIR (Failure Detection, Isolation, and Recovery)

4. Specialty Engineering
   a. EMI/EMC
   b. Reliability, maintainability, affordability, other "ilities"
   c. Logistics
   d. Etc.
5. Testability Evaluation
6. Producibility Evaluation
7. Development Testing
   a. Identification
   b. Status
   c. Results
8. Test Planning
   a. Test plan tree
   b. Status

C. *System-Level Trade Analyses*

1. Review of Key System Trades
   a. Listing of key trades
   b. Trade criteria
   c. Sensitivity analyses
   d. Change impact: Are previous trade selections still valid?

II. Second Major Design Review

- Review Action Items from previous review
- Review any relevant changes since last review

A. *Subsystem-Level Requirements Development*

1. System-to-Subsystem Specification Correlation Matrix — Confirm Traceability
   a. Identification of all TBDs, TBSs, and TBRs and assessment of criticality
   b. Identification of requirements ambiguities
2. Identification of all other requirements and constraints impinging on the subsystem
   a. Technical
   b. Cost
   c. Schedule
3. Verification Matrix identifying verification method and location in subsystem build-up
4. Definition of Subsystem Context
   a. Identification and characterization of all external interfaces by program phase

      1) High rate data
      2) Low rate data
      3) RF signals
      4) Test and diagnostic interfaces
      5) Timing, sync signals
      6) Primary and redundant power
      7) Mechanical and thermal interfaces
      8) Etc.
   b. Definition of all environments by phase
   c. Identification of all critical events by phase
   d. Identification of all modes and states by phase
   e. Operations Concept
      1) Logistics plan
      2) Maintenance plan
      3) Operability plan
      4) Etc.
   f. Mission timeline
5. Subsystem Specification Tree
6. Subsystem Plan Tree
7. Functional Definition of the Subsystem Architecture (used to correlate system-required functionality and performance with new and heritage designs implemented)
   a. Functional block diagram for each program phase
      1) Functions required with associated performance traced from system spec.
      2) Inputs
      3) Outputs
      4) Noise sources
      5) Control functions identified

B. *Synthesis*

Subsystem Design

1. Parametric Analyses Leading to Baseline Architecture
2. Subsystem Configuration Description
   a. Subsystem block diagram — Electrical architecture
   b. Subsystem layout/drawing(s) — Mechanical architecture
   c. Subsystem drawing tree
   d. Subsystem ICD
3. Identification of New Technologies Implemented
   a. Risk assessment and mitigation approaches
   b. Tailoring required for system application
4. Identification of Heritage Elements Implemented
   a. Internal interfaces identified and characterized by phase
   b. Changes required to meet subsystem application

5. Subsystem Budgets
    a. Technical: mass, power, baseband, RF, memory, throughput, etc.
    b. Cost and schedule
    c. Present margin, contingency, and reserve and location in the system
6. Subsystem Specifications
    a. Traced from system specification
    b. Listing of TBDs, TBSs, TBRs
    c. Identification of all other known requirements issues
    d. Allocation of functionality, performance, cost, schedule
7. Identification and Characterization of Internal Interfaces
    a. High rate data
    b. Low rate data
    c. RF signals
    d. Test and diagnostic interfaces
    e. Timing, sync signals
    f. Primary and redundant power
    g. Mechanical and thermal interfaces
    h. Etc.
8. Subsystem Risk Analysis
    a. Identification — Internally and externally driven
    b. Assessment of likelihood and potential impacts to system
    c. Mitigation approach(es)
9. Configuration control system
    a. Configuration control board
    b. Action item status
    c. Change notice status
    d. Subsystem database status: wire lists, budgets, etc.
    e. Interface Control
        1) Internal
        2) External
        3) ICD status
10. Operations Concept
11. Subsystem Optimization
    a. Sensitivity analyses — How sensitive is the system to potential changes?
    b. Requirements impact assessment

Subsystem Verification

1. Subsystem Simulation
    a. Correlation to specification
    b. Margin
    c. Etc.
2. FMECA (Failure Modes Effects Criticality Analysis)
3. FDIR (Failure Detection, Isolation, and Recovery)

4. Specialty Engineering
   a. EMI/EMC
   b. Reliability, maintainability, affordability, other "ilities"
   c. Logistics
   d. Etc.
5. Testability Evaluation
6. Producibility Evaluation
7. Development Testing
   a. Identification
   b. Status
   c. Results
8. Test Planning
   a. Test plan tree
   b. Status

C. Subsystem-Level Trade Analyses

1. Review of Key System Trades
   a. Listing of key trades
   b. Trade criteria
   c. Sensitivity analyses
   d. Change impact: Are previous trade selections still valid?

III. Third Major Design Review

- Review Action Items from previous review
- Review any relevant changes since last review

A. Sub-Subsystem Level Requirements Development

1. Subsystem-to-Sub-Subsystem Specification Correlation Matrix — Confirm Traceability
   a. Identification of all TBDs, TBSs, and TBRs and assessment of criticality
   b. Identification of requirements ambiguities
2. Identification of all other requirements and constraints impinging on the Sub-Subsystem
   a. Technical
   b. Cost
   c. Schedule
3. Verification Matrix identifying verification method and location in Sub-Subsystem build-up
4. Definition of Sub-Subsystem Context
   a. Identification and characterization of all external interfaces by program phase

      1) High rate data
      2) Low rate data
      3) RF signals
      4) Test and diagnostic interfaces
      5) Timing, sync signals
      6) Primary and redundant power
      7) Mechanical and thermal interfaces
      8) Etc.
   b. Definition of all environments by phase
   c. Identification of all critical events by phase
   d. Identification of all modes and states by phase
   e. Operations Concept
      1) Logistics plan
      2) Maintenance plan
      3) Operability plan
      4) Etc.
   f. Mission timeline
5. Sub-Subsystem Specification Tree
6. Sub-Subsystem Plan Tree
7. Functional Definition of the Sub-Subsystem Architecture (used to correlate sub-subsystem required functionality and performance with new and heritage designs implemented)
   a. Functional Block Diagram for Each Program Phase
      1) Functions required with associated performance traced from sub-subsystem spec.
      2) Inputs
      3) Outputs
      4) Noise sources
      5) Control functions identified

B. *Synthesis*

Sub-Subsystem Design

1. Parametric Analyses Leading to Baseline Design
2. Sub-Subsystem Configuration Description
   a. Sub-subsystem block diagram — Electrical design
   b. Sub-subsystem layout/drawing(s) — Mechanical design
   c. Sub-subsystem drawing tree
   d. Sub-subsystem ICD tree
3. Identification of New Technologies Implemented
   a. Risk assessment and mitigation approaches
   b. Tailoring required for sub-subsystem application
4. Identification of Heritage Elements Implemented
   a. Internal interfaces identified and characterized by phase
   b. Changes required to meet sub-subsystem application

5. Sub-Subsystem Budgets
   a. Technical: mass, power, baseband, RF, memory, throughput, etc.
   b. Cost and schedule
   c. Present margin, contingency, and reserve and location in the sub-subsystem
6. Lower Level Specifications (if required)
   a. Traced from subsystem specification
   b. Listing of TBDs, TBSs, TBRs
   c. Identification of all other known requirements issues
   d. Allocation of functionality, performance, cost, schedule
7. Identification and Characterization of Internal Interfaces
   a. High rate data
   b. Low rate data
   c. RF signals
   d. Test and diagnostic interfaces
   e. Timing, sync signals
   f. Primary and redundant power
   g. Mechanical and thermal interfaces
   h. Etc.
8. Sub-Subsystem Risk Analysis
   a. Identification — Internally and externally driven
   b. Assessment of likelihood and potential impacts to sub-subsystem
   c. Mitigation approach(es)
9. Configuration Control System
   a. Configuration control board
   b. Action item status
   c. Change notice status
   d. Sub-Subsystem database status: wire lists, budgets, etc.
   e. Interface control
      1) Internal
      2) External
      3) ICD Status
10. Operations Concept
11. Sub-Subsystem Optimization
    a. Sensitivity analyses — How sensitive is the sub-subsystem to potential changes?
    b. Requirements impact assessment

Sub-Subsystem Verification

1. Sub-Subsystem Simulation
   a. Correlation to specification
   b. Margin
   c. Etc.
2. FMECA (Failure Modes Effects Criticality Analysis)
3. FDIR (Failure Detection, Isolation, and Recovery)

4. Specialty Engineering
   a. EMI/EMC
   b. Reliability, maintainability, affordability, other "ilities"
   c. Logistics
   d. Etc.
5. Testability Evaluation
6. Producibility Evaluation
7. Development Testing
   a. Identification
   b. Status
   c. Results
8. Test Planning
   a. Test plan tree
   b. Status

C. *Sub-Subsystem-Level Trade Analyses*

1. Review of Key Sub-Subsystem Trades
   a. Listing of key trades
   b. Trade criteria
   c. Sensitivity analyses
   d. Change impact: Are previous trade selections still valid?

appendix D

A SDF-Derived Curriculum

This appendix outlines a framework for a System Design and Management Curriculum with suggested course content based upon the SDF. As discussed in the preceding chapters, all technical activities of the SDF can be organized into one of the basic categories: Requirements Development, Synthesis (Design, Analysis, Integration, and Verification) and Trade Analysis. This suggests a curriculum whose core technical subjects include these key topics. Managerial activities include topics such as: risk management, configuration management, subcontracts management, program planning, man-power planning, metrics definition and implementation, business development, customer interface, etc. This suggests a set of core System and Project Management courses that cover these topics in some detail. This curriculum is focused on the system development phase for complex system development.

Section 1
SDF-Derived Curriculum — Technical

I. Overall System Development Framework

Integrate into Requirements Development and Synthesis Courses

A. Time Domain View

B. Logical Domain View

C. Survey existing Standards

II. Requirements Development — Full Semester Course

B. Work Generation Activities

1. Derive Context Requirements

- QFD
- Requirements traceability
- Requirements traceability tools (doors, RTM, etc.)
- Requirements/specification writing
- Operations concept
- Program timeline
- Mission phases, modes, and states
- External interfaces

2. Generate Functional Description

- Function identification and diagramming
- Function interaction/interfaces

- Simulation of functional description
- Tools

C. *Rework Discovery Activities*

1. *Analyze Requirements*

- Implementation of Configuration Management activities
- Identification of all "To Be Determined" (TBD) holes in the requirements with a closure plan
- Identification of conflicting or inconsistent requirements with a closure plan
- Interpretation of vague or ambiguous requirements in order to review with the customer and gain his/her consensus
- Determination of the verification method (test, analysis, demonstration, simulation, inspection) that will be used for each requirement
- Determination of where in the system build-up each requirement will be verified

2. *Analyze Functional Description*

- Determine if the specification(s) is (are) complete and self-consistent
- Identify all functional requirements flowing out of imposed and derived requirements
- Determine performance requirements of each function and the relationships (interfaces, interdependencies, etc.) between functions
- Validated specification(s)
- Functional models (block diagrams, flow diagrams, behavior diagrams, simulations)

III. Synthesis — Full Semester Course

A. *Work Generation Activities*

1. *Design*

- Design space definition — parametric methods
- Partitioning — DSM
- New technology development
- CAE, CAD, CAM tools

2. *Allocation*

- Margin and contingency rules (e.g., NASA and DOD guidelines)
- Technical budget development and management

- Schedule development and management
- Cost estimation and management
- Technical performance measures, metrics

3. *Analysis*

- Form and function
- Partitioning — DSM techniques
- Lower level specification development
- Tools (RDD-100, etc.)

4. *Functional Decomposition*

5. *Inter-Tier Interaction*

- Data quality and control of information flow down
- Interface control/management

6. *Design Integration*

- Interface identification and characterization
  - Data (buses, discrete, serial, parallel, protocols)
  - Power (primary, secondary, RF, optical, etc.)
- Risk, configuration, and subcontract management (system and project management course)

B. *Rework Discovery Activities*

1. *Design Phase Test Activities*
   a. Test planning
   b. Analyses
   c. Testing engineering test models (ETM), breadboards, etc.

2. *Testability, Producibility, Specialty Engineering Analyses*

3. *Optimization*

- Linear, non-linear, integer programming techniques (system optimization course)
- Other techniques

IV. Trade Analyses

Discuss the various trade-off methodologies in the literature (cf. Footnote 57, page 85, where four sources are identified).

Section 2
SDF-Derived Curriculum — Managerial

I. Content of System Development Management Courses

A. Developing the Program Structure

Part of System and Project Management Course

- Partitioning
- Interaction control logic

B. Program Management

Part of System and Project Management Course

- Risk management
- Configuration management
- Cost development and management
- Schedule development and management
- Roles and responsibilities definition
- Reviews and audits
- Subcontracts management
- Customer interface
- Senior management interface

C. General Managerial Skills — Core Courses

1. Business development and marketing

- Proposal development and management
- Red team reviews
- Business plan development

- New technology development and management
- Short and long-term business strategy
- Technical presentations
- Marketing and customer strategy

2. *Accounting and Finance*

3. *System Dynamics*

4. *Engineering Risk Benefit Analysis*

D. Functional Management

Part of Organizational Processes Course

- Team dynamics
- Hiring practices
- Benefits, compensation, incentives
- Performance appraisals

II. Manufacturing, Production, and Distribution

Operations Management Course

Notional SDF-Based Assignments

I. Develop Requirements

Activities

- Collect and analyze imposed requirements
- Derive requirements through mission analysis, functional analysis, design, allocation, and decomposition
- Manage requirements derived during the development process
- Communicate requirements and requirements changes
- Determine and track how and where in the system build-up the requirements will be verified
- Maintain traceability of requirements
- Change impact analysis

A. Inputs

Collect Inputs

- Immediate customer
- Subcontractors
- Heritage designs
- The department
- The division
- Procuring organization
- New technology
- Competitors
- Business development
- The corporation
- User community

B. Work Generation Activities

1. Derive Context Requirements

Activities/Assignments:

Identify all mission phases, modes, and states
Identify and characterize all external interfaces, by mission phase
Define the environments to which the system will be subjected, by mission phase
Identify critical issues by mission phase (events, technologies, etc.)
Develop the concept of operations

Output

- Specification(s)
- Operations concept
- Context diagram(s)
- Entity relation diagram(s)
- Event list(s)
- External ICDs
- FMECA

2. Generate Functional Description

Activities/Assignments:

Identify all functional requirements flowing out of imposed and derived requirements
Develop the specification(s)
Determine performance requirements of each function and the relationships (interfaces, interdependencies, etc.) between functions

Output

Specification(s)
Functional models (block diagrams, flow diagrams, behavior diagrams, simulations)

C. Rework Discovery Activities

1. Analyze Requirements

Activities/Assignments:

Outline implementation of Configuration Management activity

- Identify all of "To Be Determined" (TBD) holes in the requirements with a closure plan
- Identify conflicting or inconsistent requirements with a closure plan
- Interpret vague or ambiguous requirements in order to review with the customer and gain consensus
- Determine the verification method (test, analysis, demonstration, simulation, inspection) that will be used for each requirement
- Determine where in the system build-up each requirement will be verified

2. *Analyze Functional Description*

Activities/Assignments:

- Determine if the specification(s) is (are) complete and self-consistent
- Identify all functional requirements flowing out of imposed and derived requirements
- Determine performance requirements of each function and the relationships (interfaces, interdependencies, etc.) between functions

Output

Validated specification(s)
Functional models (block diagrams, flow diagrams, behavior diagrams, simulations

Output → Functional Description
Customer Consensus — Gain Instructor Consensus

II. Synthesis

A. Work Generation Activities

1. *Design*

Activities/Assignments:

- Quantify Design Space (H/W & S/W)
    - Parametric Analyses
    - New technologies and heritage designs are surveyed for applicability
- Generate Preliminary Design
    - Block diagrams, Schematics, Drawings, etc.
    - Internal ICDs
- Risk Management → Identify and Assess Risk
    - Technical performance, cost, schedule
    - Preliminary mitigation approaches
- Configuration Management of all design documentation

Output → H/W & S/W concept(s) and/or design(s), Risk assessment

2. *Allocation*

Activities/Assignments:

- Allocate functionality, performance, constraints to H/W and S/W elements
- Define budgets
  - Technical: mass, power, throughput, memory, RF links, etc.
  - Reliability, Contamination, etc.
  - Margin and Contingency Rules
- Performance Monitoring — Metrics Development — Define/Refine TPMs
- Cost and Schedule Management

Output → Budgets, Technical Performance Measures

3. *Analysis*

Activities/Assignments: Perform the following as appropriate:

- Mission, system, electrical, digital, analog, RF, mechanical, etc.
- Simulations
- FMECA (Failure Modes Effects and Criticality Analysis)

4. *Functional Decomposition*

Activities/Assignments:

- Decompose the implementation into subfunctions
- Identify the interfaces between the subfunctions
- Generate the functional model and verify the functional definition
- Generate the specification(s) and ICDs

Output → Lower-level validated specifications and ICD(s), lower-level simulation

5. *Integration*

Activities/Assignments:

- Identify and characterize interfaces
- Update Specifications, ICDs, Databases, Etc.
- Update Design Definition

- Update Mission Timeline and Operations Concept
- Develop Block diagrams, schematics, Drawings, layouts
- Management Activities
  - Performance Measurement — Budgets, etc.
  - Subcontract Management
  - Risk Management — Identification, assessment, and mitigation approaches
  - Configuration Management — Configuration Control Board (CCB)

Output → Integrated design that includes the data generated above

B. Rework Discovery Activities

Activities/Assignments:

1. Analysis, Development Testing, and Test Planning

- Analysis — Perform those analyses aimed at determining "how well" the current design meets its requirements. This is in contrast to those analyses aimed at defining design space which are performed as described above in the design activity.
- Perform development testing as appropriate (e.g., ETMs, prototypes, breadboards)
- Develop product test plans (e.g., test requirements, test flow, resource planning, etc.)

2. Producibility, Testability, and Other Specialty Engineering Activities

These activities assess those areas of the design commonly called "specialty engineering" concerns.

- Assess testability within resource and time constraints
- Assess producibility within resource and time constraints
- Assess acceptability with respect to EMI/EMC, reliability, maintainability, affordability, supportability, etc. parameters

3. Optimization

Discuss various optimization approaches. Perform appropriate optimization analyses.

Output → The output of the Design Verification activity is a design that has been assessed as to how well it meets all the requirements.

III. Trade Analysis

Discuss the pros and cons of the various trade methodologies found in the literature. Perform a trade study commensurate with customer requirements, program need, and other criteria as determined by the development team.

Highlight the classic trade-off that occurs in most any system development activity: cost, schedule, and technical performance. Also include issues relating to risk, robustness, safety, cost, schedule, and technical performance.

appendix E

Mapping EQFD and Robust Design into the SDF

For those interested in applying Robust Design and Quality Function Deployment (QFD) techniques into their engineering processes, the basic steps of each have been mapped into the SDF.[69] This mapping is not to be taken as absolute, but as a guide as to how those activities might be integrated into the overall framework.

I. Requirements Development

A. Inputs

B. Work Generation Activities

1. Derive Context Requirements

- Identify all mission phases, modes, and states
- Develop mission timeline and operations concept
- Identify all external interfaces by mission phase
- Identify critical issues by mission phase (events, technologies, etc.)
- Define environments by mission phase
- Room 2: Context definition: Corporation
  - Analyze full life cycle (EQFD 8-9)[70]
  - Preliminary noise source identification

2. Generate Functional Description

Develop and validate the functional description of the system. Determine if the specification(s) is (are) complete and self-consistent.

[69] The EQFD activities were drawn from Clausing, Don, "Total Quality Development," ASME Press, 1994. The Robust Design activities were drawn from Madhav S. Phadke, "Quality Engineering Using Robust Design," Prentice Hall, 1989.
[70] EQFD is Enhanced Quality Function Deployment, as defined by Clausing.

- Identify all functional requirements flowing out of imposed and derived requirements
- Determine performance requirements of each function and the relationships (interfaces, interdependencies, etc.) between functions
- Tool for specification validation by simulation
- Define preliminary ideal function — signal-to-noise

C. Rework Discovery Activities

1. Analyze Requirements

Determine fidelity of the input or imposed requirements.

- Room 1: Requirements capture: Customer (EQFD 1-7)
  - Identify all customer and user requirements
  - Analyze for completeness, consistency, conflicts, etc.

Determine if all imposed and derived requirements are verifiable.

- Determine where in the system build-up each requirement will be verified
- Determine Verification method (test, analysis, demonstration, simulation, inspection)
- Verifiability (EQFD 10)
- Analyze value & Rooms 3, 6, 7: EQFD customer analyses
- Rooms 4 & 5 benchmarking — see Section II
- Room 8: Alternatives and priority definition — See Section II

2. Analyze Functional Description

- Determine if the specification(s) is (are) complete and self-consistent
- Identify all functional requirements flowing out of imposed and derived requirements
- Determine performance requirements of each function and the relationships (interfaces, interdependencies, etc.) between functions

II. Synthesis

A. Work Generation Activities

1. Design

- Parametric analyses
- New technologies and heritage designs are surveyed for applicability
- Identify benchmark products
- Customer competitive survey (EQFD 13, Room 4)

- Corporate measures (EQFD 14, Room 5)
- Orthogonal arrays — variation, value-of-performance
- Select critical design parameters: Fault trees, etc.
- Define feasible alternate design values (Product Parameters Design Step 2; EQFD 19, Room 8)
- Select areas for concentrated effort (Product Parameter Design Step 3; EQFD 20, Room 8)
- Block diagrams, schematics, drawings, etc.
- Tolerance design — quality loss function, precision vs. cost vs. quality loss
- Process parameter design
- Technical performance, cost, schedule
- Preliminary mitigation approaches

2. *Allocation*

   a. Allocate: functionality, performance, constraints, etc. to system H/W and S/W elements
   b. Define preliminary budgets

- Technical: mass, power, throughput, memory, RF links, etc.
- Cost
- Schedule
- Risk, reliability, contamination, etc.

   c. Margin and contingency rules

- Define rules
- Implement in Budgets

   d. Define/refine technical performance measures (TPM)

3. *Analysis*

a. Mission, system, electrical, digital, analog, RF, mechanical, etc.
b. Simulations
c. FMECA (Failure Modes Effects and Criticality Analysis)

4. *Functional Decomposition*

- Decompose in single level increments
- Identify interfaces between next-tier elements
- Generate functional model
- Generate lower-level specification(s)
- Validate lower level specs
- Generate interface control document(s)

5. Integration

e. Update design definition

- Update mission timeline
- Operations concept
- Block diagrams, schematics, drawings, layouts
- Identify and characterize interfaces
- Budgets, etc.
- Risk management — update risk identification, assessment, and mitigation approaches

f. Configuration management

- Configuration control board (CCB)
- Specifications
- Interface control document(s)
- Databases, etc.

g. QFD and Robust Development

- Refine ideal function definition
- Refine critical parameter design, tolerance design, process parameter design, etc.

B. Rework Discovery Activities

1. Analyses, Development Testing, Test Planning

These involve any and all analyses necessary to ascertain technical, cost, schedule, robustness, and risk performance of the system. Note that the "Impose Noise and Evaluate" activity is considered an essential element of the Engineering Technical Process.

- Mission, system, electrical, digital, analog, RF, mechanical, etc.
- Impose noise and evaluate (Product Parameters Design Step 4 & 5)
- Analyze benchmark products:
  - Customer competitive survey (EQFD 13, Room 4)
  - Corporate measures (EQFD 14, Room 5)

2. Producibility, Testability, and Other Specialty Engineering Activities

- Is the design testable within resource and time constraints?
- Is the design producible within resource and time constraints?

- Is the design acceptable with respect to EMI/EMC, reliability, maintainability, affordability, supportability, etc. parameters?
- Evaluate technical difficulty (EQFD 18, Room 7)

III. Trade Analyses

The definition of which specific activities should be included under the broad heading of "trade analyses" varies in the literature. The SDF places trade analyses at the end of the process (recall discussion in Chapter 5, page 84), because cost, schedule, technical, and risk parameters must be developed in order to make a selection. Thus, while many elements of QFD and robust design impact trade analyses, none are included under this heading since they occurred previously.

appendix F

A Simple System Dynamics Model of the SDF

This System Dynamics model was developed using Vensim® Professional32 Version 3.0C. It is provided as a reference for anyone interested in pursuing this more fully. There are many excellent texts on the subject and the interested reader is referred to those works.

(01) Cum SS Tasks = INTEG (SS Tasks Gen'd, 0)
Units: Task
Uses: (20)SS Finished
(25)SS Task Gen

(02) Cum Sys Syn Tasks = INTEG (Sys Syn Tasks Gen'd, 0)
Units: Task
Uses: (46)Sys Syn Finished
(52)Sys Syn Task Gen
(53)Sys Syn Tasks Gen'd

(03) Cum UD RW= INTEG (Cum UD RW Rate, 0)
Units: Task

(04) Cum UD RW Rate = Sys Rqmt RW Gen + Sys Syn RW Gen + SS RW Gen
Units: Task/Month
Uses: (03)Cum UD RW

(05) FINAL TIME = 100
Units: Month

(06) Handoff Constraint = 0
Units: Dimensionless
Uses: (26)SS Task Released
(55)Sys Syn Tasks Released

(07) INITIAL TIME = 0
Units: Month

(08) Nominal Quality = 0.5
Units: Dimensionless
Uses: (22)SS Quality
(39)Sys Rqmt Quality
(49)Sys Syn Quality

(09) Number of SS = 1
Units: Dimensionless
Uses: (25)SS Task Gen
(35)Syn Tasks for SS

(10) Productivity = 0.5
Units: Task/(Man * Month)
Uses: (13)RW Pot Work Rate
(21)SS Potential Work Rate
(38)Sys Rqmt Potential Work Rate
(47)Sys Syn Potential Work Rate

(11) Release Delay = 1
Units: Month
Uses: (26)SS Task Released
(55)Sys Syn Tasks Released

(12) RW Percent = 0.125
Units: Dimensionless
Uses: (14)RW Staff Level

(13) RW Pot Work Rate = Productivity * RW Staff Level
Units: Task/Month
Uses: (23)SS RW Disc
(40)Sys Rqmt RW Disc
(50)Sys Syn RW Disc

(14) RW Staff Level = Staff Level * RW Percent
Units: Man
Uses: (13)RW Pot Work Rate

(15) SAVEPER = TIME STEP
Units: Month

(16) SR Prcv'd Fraction Complete = (SR Work Prcv'd Complete)/Tasks To Do
Units: Dimensionless
Uses: (52)Sys Syn Task Gen
(55)Sys Syn Tasks Released

(17) SR Tasks for Sys Syn = Tasks To Do * 1
Units: Task
Uses: (52)Sys Syn Task Gen
(53)Sys Syn Tasks Gen'd

(18) SR Work Prcv'd Complete = Sys Rqmt Undisc RW + Sys Rqmt Work Done
Units: Task
Uses: (16)SR Prcv'd Fraction Complete

(19) SRD Avg Qlty = Sys Rqmt Work Done / max (0.001, Sys Rqmt Work Done + Sys Rqmt Undisc RW)
Units: Dimensionless
Uses: (49)Sys Syn Quality

(20) SS Finished = IF THEN ELSE (SS Work Done / max (0.001, Cum SS Tasks) > 0.9999, 0, 1)
Units: Dimensionless
Uses: (23)SS RW Disc
(24)SS RW Gen
(30)SS Work Acc

(21) SS Potential Work Rate = Productivity * Staff Level
Units: Task/Month
Uses: (24)SS RW Gen
(30)SS Work Acc

(22) SS Quality = Nominal Quality * Syn Avg Qlty
Units: Dimensionless
Uses: (24)SS RW Gen
(30)SS Work Acc

(23) SS RW Disc = IF THEN ELSE (SS Undisc RW<= 0, 0, min (RW Pot Work Rate, RW Pot Work Rate * SS Undisc RW)) * SS Finished
Units: Task/Month
Uses: (29)SS Undisc RW
(32)SS Work To Do

(24) SS RW Gen = IF THEN ELSE (SS Work To Do <= 0, 0, 1) * min (SS Work To Do, SS Potential Work Rate) * (1 – SS Quality) * SS Finished
Units: Task/Month
Uses: (29)SS Undisc RW
(32)SS Work To Do
(04)Cum UD RW Rate

(25) SS Task Gen = IF THEN ELSE (Cum SS Tasks > Syn Tasks for SS * Table for SS Tasks wrt Syn FC (Sys Syn Prcv'd Fraction Complete), 0, Number of SS * (Sys Syn Work Acc + Sys Syn RW Gen))

Units: Task/Month
Uses: (28)SS Tasks Ready
(27)SS Tasks Gen'd

(26) SS Task Released = IF THEN ELSE (Sys Syn Prcv'd Fraction Complete < Handoff Constraint, 0, 1) * SS Tasks Ready/Release Delay
Units: Task/Month
Uses: (28)SS Tasks Ready
(32)SS Work To Do

(27) SS Tasks Gen'd = SS Task Gen
Units: Task/Month
Uses: (01)Cum SS Tasks

(28) SS Tasks Ready = INTEG (SS Task Gen – SS Task Released, 0)
Units: Task
Uses: (26)SS Task Released

(29) SS Undisc RW = INTEG (SS RW Gen – SS RW Disc, 0)
Units: Task
Uses: (23)SS RW Disc

(30) SS Work Acc = IF THEN ELSE (SS Work To Do <= 0, 0, 1) * min (SS Work To Do, SS Potential Work Rate) * SS Quality * SS Finished
Units: Task/Month
Uses: (31)SS Work Done
(32)SS Work To Do

(31) SS Work Done= INTEG (SS Work Acc, 0)
Units: Task
Uses: (20)SS Finished

(32) SS Work To Do= INTEG (SS Task Released + SS RW Disc – SS Work Acc – SS RW Gen, 0)
Units: Task
Uses: (24)SS RW Gen
(30)SS Work Acc

(33) Staff Level = 5
Units: Man
Uses: (14)RW Staff Level
(21)SS Potential Work Rate
(38)Sys Rqmt Potential Work Rate
(47)Sys Syn Potential Work Rate

(34) Syn Avg Qlty = IF THEN ELSE (Sys Syn Work Done > 0, Sys Syn Work Done/max (0.0001, Sys Syn Work Done + Sys Syn Undisc RW), 1)
Units: Dimensionless
Uses: (22)SS Quality

(35) Syn Tasks for SS = Tasks To Do * Number of SS
Units: Task
Uses: (25)SS Task Gen

(36) Syn Work Prcv'd Complete = Sys Syn Undisc RW + Sys Syn Work Done
Units: Task
Uses: (48)Sys Syn Prcv'd Fraction Complete

(37) Sys Rqmt Finished = IF THEN ELSE ((Sys Rqmt Work Done/Tasks To Do) >= 0.9999, 0, 1)
Units: Dimensionless
Uses: (40)Sys Rqmt RW Disc
(41)Sys Rqmt RW Gen
(43)Sys Rqmt Work Acc

(38) Sys Rqmt Potential Work Rate = Productivity * Staff Level
Units: Task/Month
Uses: (41)Sys Rqmt RW Gen
(43)Sys Rqmt Work Acc

(39) Sys Rqmt Quality = Nominal Quality
Units: Dimensionless
Uses: (41)Sys Rqmt RW Gen
(43)Sys Rqmt Work Acc

(40) Sys Rqmt RW Disc = IF THEN ELSE (Sys Rqmt Undisc RW<= 0, 0, min (RW Pot Work Rate, Sys Rqmt Undisc RW * RW Pot Work Rate)) * Sys Rqmt Finished
Units: Task/Month
Uses: (42)Sys Rqmt Undisc RW
(45)Sys Rqmt Work To Do

(41) Sys Rqmt RW Gen = IF THEN ELSE (Sys Rqmt Work To Do <= 0, 0, 1) * min (Sys Rqmt Work To Do, Sys Rqmt Potential Work Rate) * (1 – Sys Rqmt Quality) * Sys Rqmt Finished
Units: Task/Month
Uses: (42)Sys Rqmt Undisc RW
(45)Sys Rqmt Work To Do
(04)Cum UD RW Rate
(52)Sys Syn Task Gen

(42) Sys Rqmt Undisc RW = INTEG (Sys Rqmt RW Gen – Sys Rqmt RW Disc, 0)
Units: Task
Uses: (18)SR Work Prcv'd Complete
(19)SRD Avg Qlty
(40)Sys Rqmt RW Disc

(43) Sys Rqmt Work Acc = IF THEN ELSE (Sys Rqmt Work To Do <= 0, 0, 1) * min (Sys Rqmt Work To Do, Sys Rqmt Potential Work Rate) * Sys Rqmt Quality * Sys Rqmt Finished
Units: Task/Month
Uses: (44)Sys Rqmt Work Done
(45)Sys Rqmt Work To Do
(52)Sys Syn Task Gen

(44) Sys Rqmt Work Done= INTEG (Sys Rqmt Work Acc, 0)
Units: Task
Uses: (18)SR Work Prcv'd Complete
(19)SRD Avg Qlty
(37)Sys Rqmt Finished

(45) Sys Rqmt Work To Do = INTEG (Sys Rqmt RW Disc – Sys Rqmt RW Gen – Sys Rqmt Work Acc, Tasks To Do)
Units: Task
Uses: (41)Sys Rqmt RW Gen
(43)Sys Rqmt Work Acc

(46) Sys Syn Finished = IF THEN ELSE (Sys Syn Work Done / max (0.001, Cum Sys Syn Tasks) > 0.9999, 0, 1)
Units: Dimensionless
Uses: (50)Sys Syn RW Disc
(51)Sys Syn RW Gen
(57)Sys Syn Work Acc

(47) Sys Syn Potential Work Rate = Productivity * Staff Level
Units: Task/Month
Uses: (51)Sys Syn RW Gen
(57)Sys Syn Work Acc

(48) Sys Syn Prcv'd Fraction Complete = (Syn Work Prcv'd Complete) / (Tasks To Do)
Units: Dimensionless
Uses: (25)SS Task Gen
(26)SS Task Released

(49) Sys Syn Quality = Nominal Quality * SRD Avg Qlty
Units: Dimensionless
Uses: (51)Sys Syn RW Gen
(57)Sys Syn Work Acc

(50) Sys Syn RW Disc = IF THEN ELSE (Sys Syn Undisc RW<= 0, 0, min (RW Pot Work Rate, RW Pot Work Rate * Sys Syn Undisc RW)) * Sys Syn Finished
Units: Task/Month
Uses: (56)Sys Syn Undisc RW
(59)Sys Syn Work To Do

(51) Sys Syn RW Gen = IF THEN ELSE (Sys Syn Work To Do <= 0, 0, 1) * min (Sys Syn Work To Do, Sys Syn Potential Work Rate) * (1 – Sys Syn Quality) * Sys Syn Finished
Units: Task/Month
Uses: (56)Sys Syn Undisc RW
(59)Sys Syn Work To Do
(04)Cum UD RW Rate
(25)SS Task Gen

(52) Sys Syn Task Gen = IF THEN ELSE (Cum Sys Syn Tasks > Table Sys Syn Tasks wrt SR FC (SR Prcv'd Fraction Complete) * SR Tasks for Sys Syn, 0, Sys Rqmt Work Acc + Sys Rqmt RW Gen)
Units: Task/Month
Uses: (54)Sys Syn Tasks Ready
(53)Sys Syn Tasks Gen'd

(53) Sys Syn Tasks Gen'd = Sys Syn Task Gen * IF THEN ELSE (Cum Sys Syn Tasks >= SR Tasks for Sys Syn, 0, 1)
Units: Task/Month
Uses: (02)Cum Sys Syn Tasks

(54) Sys Syn Tasks Ready = INTEG (Sys Syn Task Gen – Sys Syn Tasks Released, 0)
Units: Task
Uses: (55)Sys Syn Tasks Released

(55) Sys Syn Tasks Released = IF THEN ELSE (SR Prcv'd Fraction Complete < Handoff Constraint, 0, 1) * Sys Syn Tasks Ready/Release Delay
Units: Task/Month
Uses: (54)Sys Syn Tasks Ready
(59)Sys Syn Work To Do

(56) Sys Syn Undisc RW = INTEG (Sys Syn RW Gen – Sys Syn RW Disc, 0)
Units: Task
Uses: (34)Syn Avg Qlty
(36)Syn Work Prcv'd Complete
(50)Sys Syn RW Disc

(57) Sys Syn Work Acc = IF THEN ELSE (Sys Syn Work To Do <= 0, 0, 1) * min (Sys Syn Work To Do, Sys Syn Potential Work Rate) * Sys Syn Quality * Sys Syn Finished
Units: Task/Month
Uses: (58)Sys Syn Work Done
(59)Sys Syn Work To Do
(25)SS Task Gen

(58) Sys Syn Work Done = INTEG (Sys Syn Work Acc, 0)
Units: Task
Uses: (34)Syn Avg Qlty
(36)Syn Work Prcv'd Complete
(46)Sys Syn Finished

(59) Sys Syn Work To Do = INTEG (Sys Syn Tasks Released + Sys Syn RW Disc – Sys Syn RW Gen – Sys Syn Work Acc, 0)
Units: Task
Uses: (51)Sys Syn RW Gen
(57)Sys Syn Work Acc

(60) Table for SS Tasks wrt Syn FC([(0,0),(1,1)],(0,0),(0.5,1))
Units: Dimensionless
Uses: (25)SS Task Gen

(61) Table Sys Syn Tasks wrt SR FC([(0,0),(1,1)],(0,0),(0.5,1))
Units: Dimensionless
Uses: (52)Sys Syn Task Gen

(62) Tasks To Do = 50
Units: Task
Uses: (45)Sys Rqmt Work To Do
(16)SR Prcv'd Fraction Complete
(17)SR Tasks for Sys Syn
(35)Syn Tasks for SS
(37)Sys Rqmt Finished
(48)Sys Syn Prcv'd Fraction Complete

(63) TIME STEP = 0.0625
Units: Month

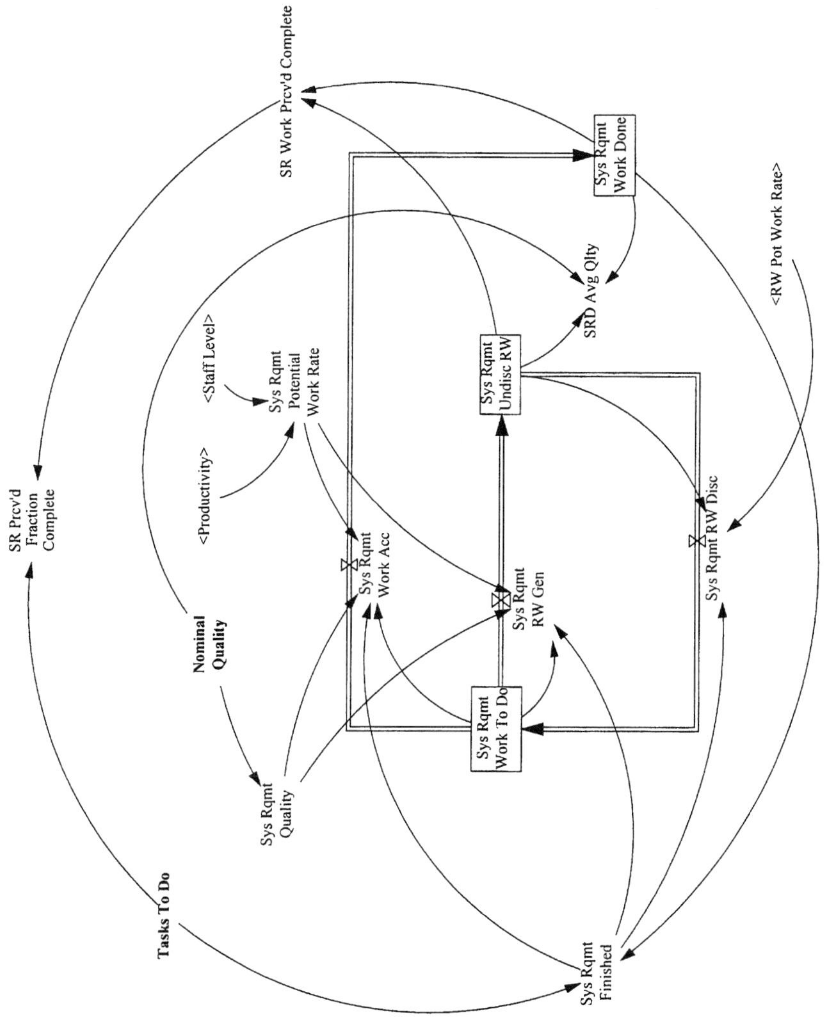

Figure F1 Requirements Development Phase.

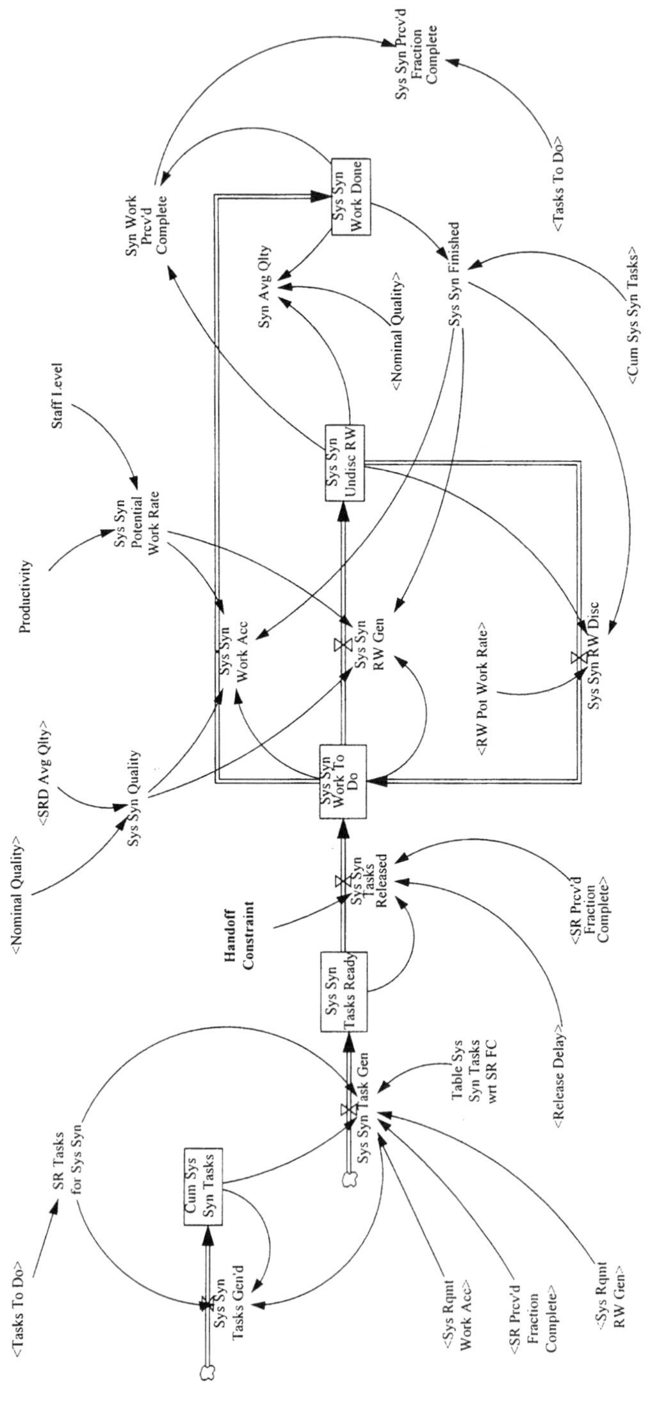

Figure F2 Synthesis Phase.

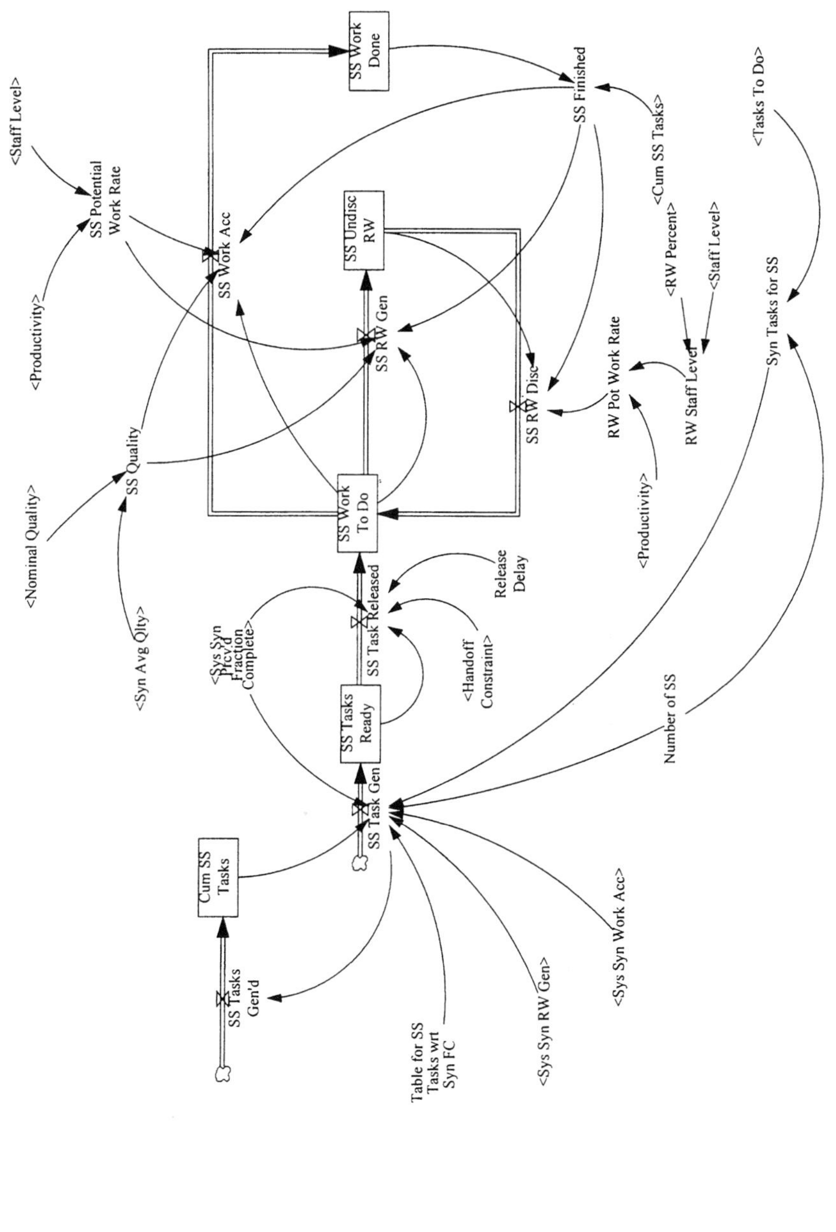

Figure F3 Subsystem Phase.

appendix G

SDF Presentation Slides

A System Design and Management Framework

 1

Key Questions in SE Process Development

- What is a system?
- What are the major activities and sub-activities that make up the system engineering process?
- What is the logical sequence of those activities?
- What are the feedbacks within the process and why do they exist?
- How do the various levels within the program hierarchy interrelate?
- How are the major activities decomposed and how do they relate to one another?
- When should iteration occur and how should it be planned for?

 2

Working Definition of "System"

"Any Entity Within Prescribed Boundaries That Performs Work on an Input in Order to Generate an Output"

 3

SE Literature Search

<u>Industry Standards</u>

- IEEE-1220
- EIA/IS-632
- Mil-Std-499A
- Army Field Manual 770-786

<u>Individual Authors</u>

- Shinners
- Reinert & Wertz
- Coutinho
- Hall
- Blanchard & Fabrycky
- Chase
- Wymore

<u>Consensus</u>

- What
- How
- How Well
- Verify
- Select

 4

SDF Basic Building Block

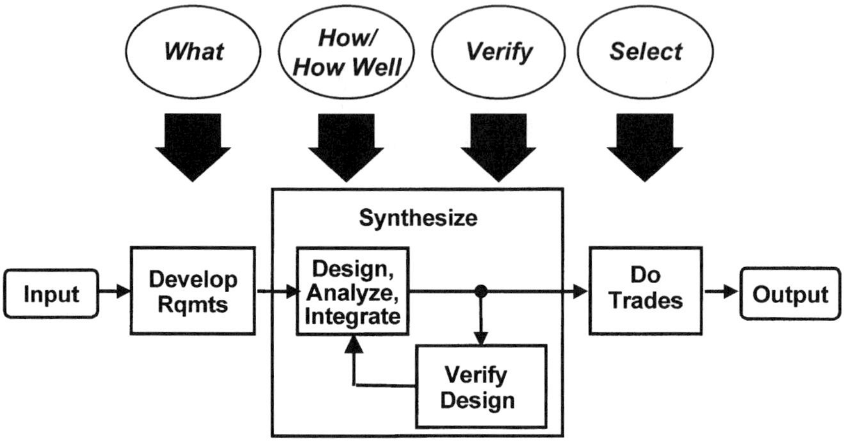

 5

The Time and Logical Domains

 6

Both Time & Logical Views Needed to Clearly Describe Program

- Why have some efforts had limited success in defining a generalized process applicable to many contexts?
  - Time and logical domains not explicitly identified and characterized in distinction
  - When the logical view is overlaid on a chronological view, the resulting process becomes application specific
- When characterized in distinction the overall framework is preserved

 7

Both Views Needed to Provide Full Program Description

	Major Milestone Reviews				
System Level	Award	First	Second	Third	Fourth
Level 0 Rqmts	Draft	Final	Update	Update	Update
Level 0 Design	Initial	Draft	Final	Update	Update
Level 1 Rqmts	Initial	Draft	Final	Update	Update
Level 1 Design	—	Initial	Draft	Final	Update
Level 2 Rqmts	—	Initial	Draft	Final	Update
Level 2 Design	—	—	Initial	Draft	Final
Level 3 Rqmts	—	—	Initial	Draft	Final
Level 3 Design	—	—	—	Initial	Draft

TIME DOMAIN

How I/O evolves over time in terms of kind/ type, fidelity, and depth or level of detail

LOGICAL DOMAIN

Instantaneous program state in terms of energy expended to perform activities and generate data

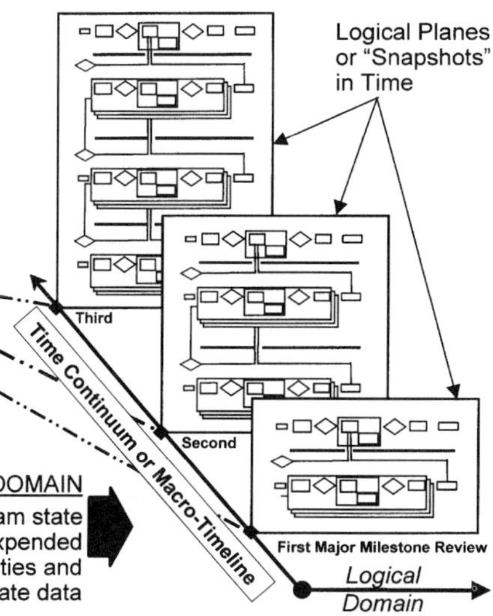

 8

Time Domain ➔ Output Progression f(time)

- Describes how output evolves over time
- Over time energy is expended in each activity until the desired output at the necessary fidelity is generated
- Two distinct sets of data are generated: Requirements and Design
  - Requirements are developed in sufficient detail such that design activities can be performed with reasonable probability of success ➔ Acceptable Risk
  - The design definition also evolves over time with increasing detail at increasingly lower levels of the system hierarchy.

 9

The SDF in the Time Domain

System Tier	Award	First	Second	●●●	"N"
Level 0 Rqmts	Draft	Final	Update	●●●	Update
Level 0 Design	Initial	Draft	Final		Update
Level 1 Rqmts	Initial	Draft	Final	●●●	Update
Level 1 Design	—	Initial	Draft		Update
Level 2 Rqmts	—	Initial	Draft	●●●	Update
Level 2 Design	—	—	Initial		Final
Level "n" Rqmts	—	—	Initial	●●●	Final
Level "n" Design	—	—	—		Draft

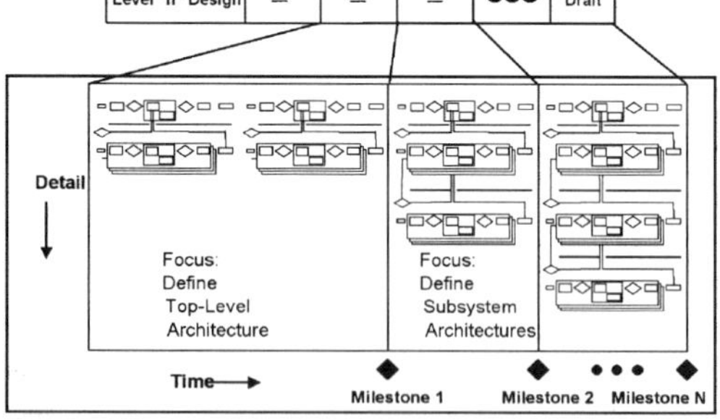

Time Domain Focus: Output as a Function of Time

 10

Logical Domain: Instantaneous Snapshot of Program State

- At any instant in time each activity is performed at some level of intensity at some tier of system hierarchy
- The level of intensity is dependent upon many factors:
  - stability of the input requirements
  - level of complexity of the system
  - whether the system is precedented or not
  - where on the timeline the development effort is occurring
- The time continuum contains an infinite number of "logical planes", each reveals:
  - How many tiers are involved and how many subsystems
  - Logical connections within and between each tier
  - Energy level applied to each activity

 11

The SDF Logical View

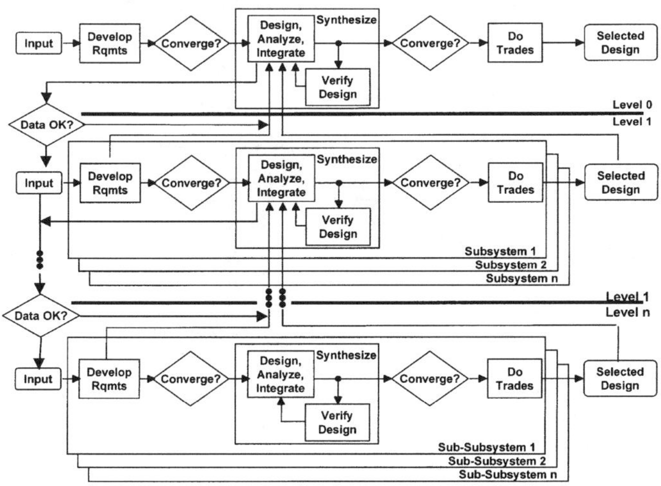

Logical Domain: "Snapshot" of Energy Expenditure

 12

The Rework Cycle

 13

Cost Nemesis—Rework

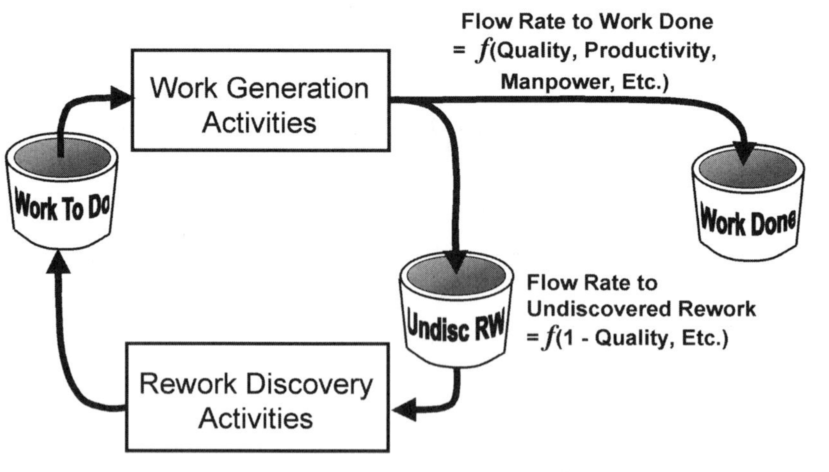

 14

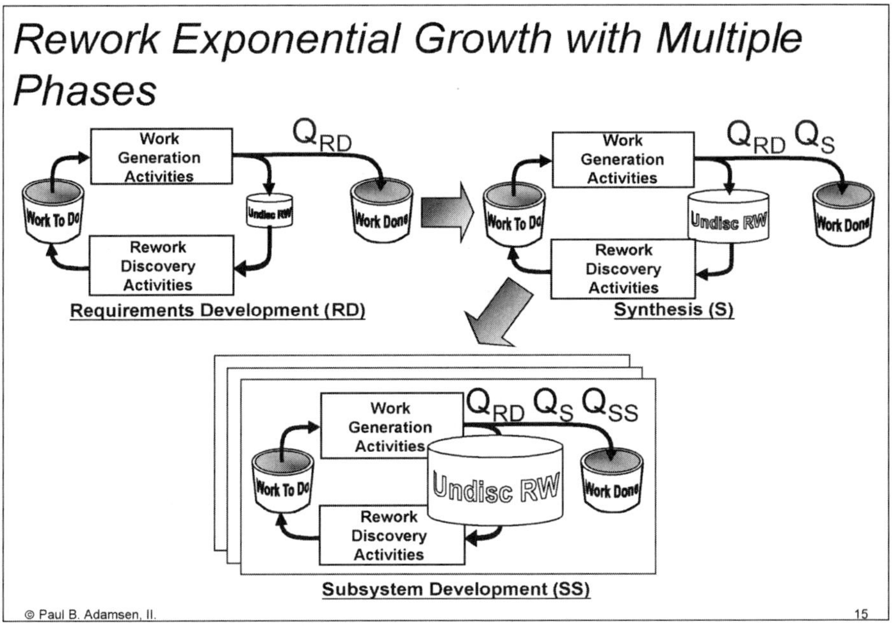
Rework Exponential Growth with Multiple Phases
Work Generation Activities
Q_{RD}
Work To Do
Undisc RW
Work Done
Rework Discovery Activities
Requirements Development (RD)
Work Generation Activities
Q_{RD} Q_S
Work To Do
Undisc RW
Work Done
Rework Discovery Activities
Synthesis (S)
Work Generation Activities
Q_{RD} Q_S Q_{SS}
Work To Do
Undisc RW
Work Done
Rework Discovery Activities
Subsystem Development (SS)
© Paul B. Adamsen, II.
15

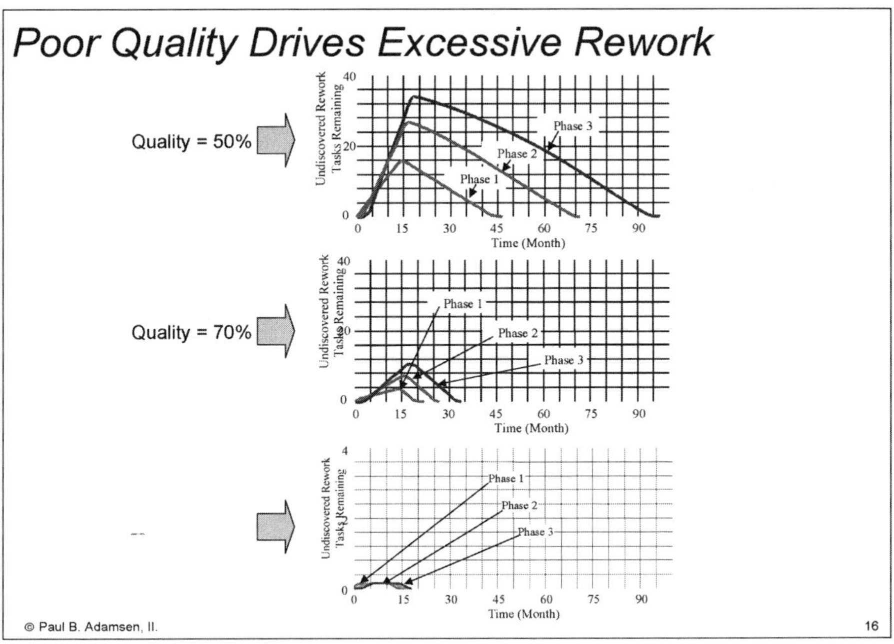
Poor Quality Drives Excessive Rework
Quality = 50%
Undiscovered Rework Tasks Remaining
Phase 3
Phase 2
Phase 1
Time (Month)
Quality = 70%
Undiscovered Rework Tasks Remaining
Phase 1
Phase 2
Phase 3
Time (Month)
Undiscovered Rework Tasks Remaining
Phase 1
Phase 2
Phase 3
Time (Month)
© Paul B. Adamsen, II.
16

Insufficient Effort Focused on Discovering Rework Leads to Still More Rework

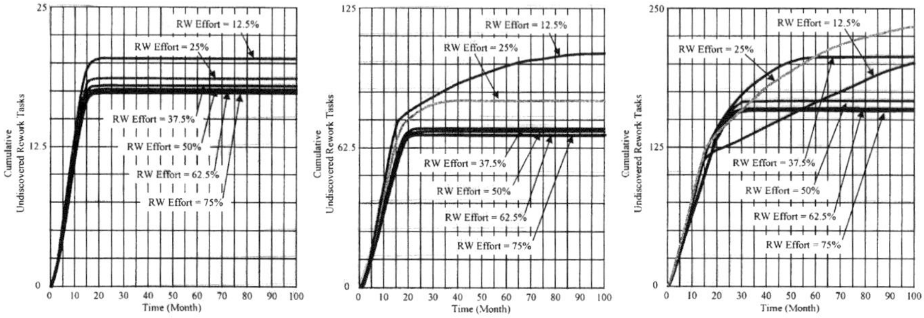

Quality=90% Quality=70% Quality=50%

 17

So What's the "So What"?

- Rework (undiscovered & known) is one of the major causes of cost and schedule overruns
- Quality is the greatest leverage point to improve program cost and schedule performance
- Rework Discovery is essential for controlling system development cost and schedule

 18

What is the Purpose of Each SDF Activity?

Requirements Development		Design & Analysis				Verification	
Activity	Main Focus	Activity	Main Focus	Activity	Main Focus	Activity	Main Focus
Requirements Analysis	Discover Rework	Identify/Modify Design	Work	Analyze Performance	Discover Rework	Analysis (may be same as Synthesis)	Discover Rework
Mission Analysis	Work	Allocation	Work	Assess Producibility, Testability	Discover Rework	Test	Discover Rework
Rqmts Verification Check	Discover Rework	Functional Decomposition	Work	Optimize	Work	Plan System Test	Discover Rework
Functional Analysis	Work & Discover Rework	Design Integration	Work				

About Half the SDF Activities Focus on Rework Discovery

The System Development Framework: Technical

SDF in the Logical Domain

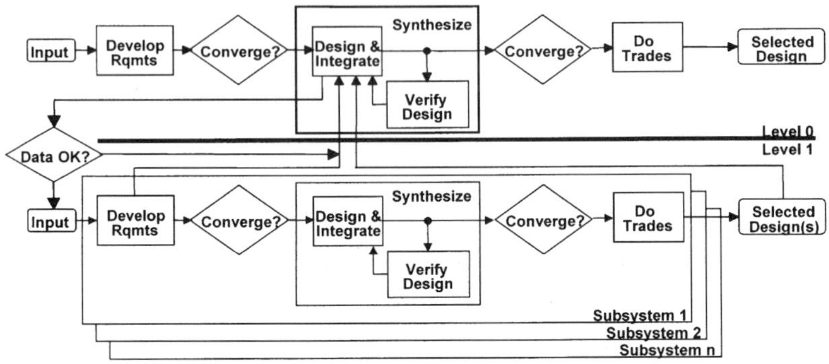

- Same process is used at each product level
- Identifies information flow paths, I/F control responsibility
- Ensures "closed-loop" development tailorable to specific program needs
- Modularity facilitates Tailoring

 21

SDF Modular Construction

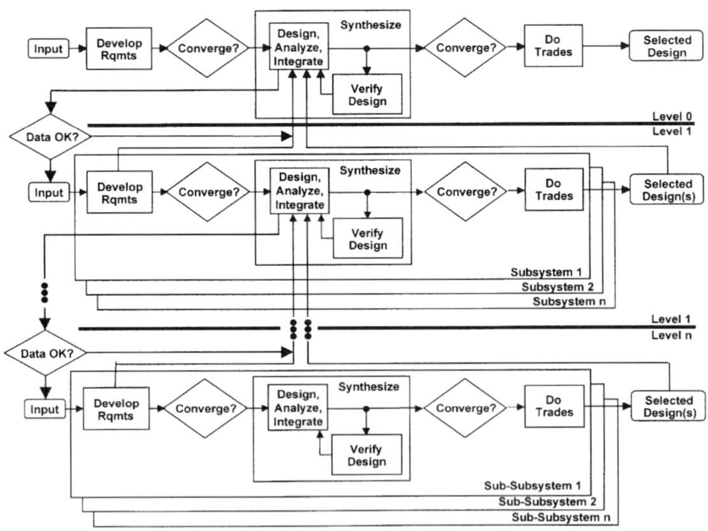

 22

Convergence

- A key criterion in moving from one activity to another is convergence
- Examples Non-Convergence
  - Spacecraft required to communicate with a relay satellite while their orbits preclude such communication
  - When a function cannot be performed without input from another function
  - Unavailability of certain technologies required to satisfy a particular requirements set
  - Stable and consistent requirements but not implementable—at a reasonable cost or schedule
- Occurs when there is an acceptable probability of success that the subsequent activity will converge with that data

 23

Requirements Development

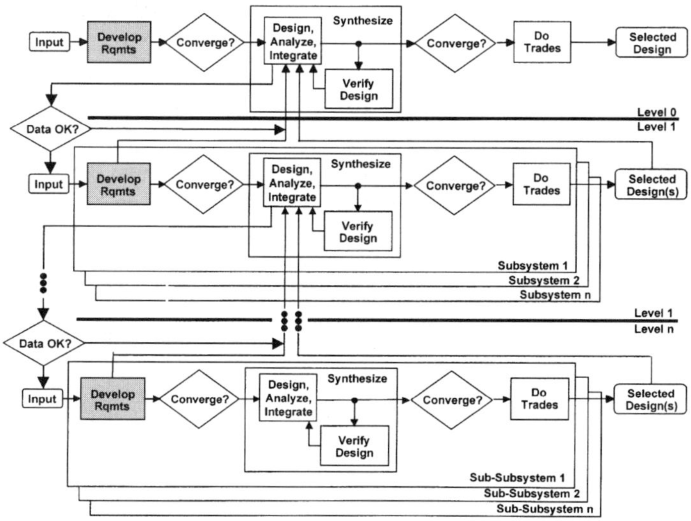

 24

"Develop Requirements" Activity Decomposed

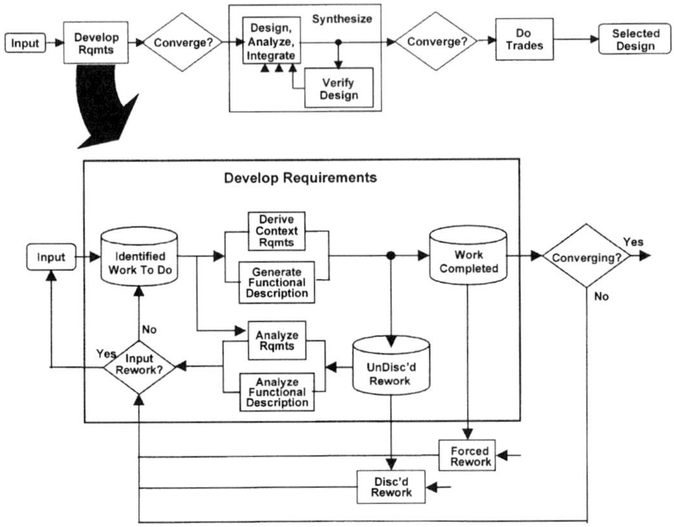

 25

Requirements Development: Work Activities

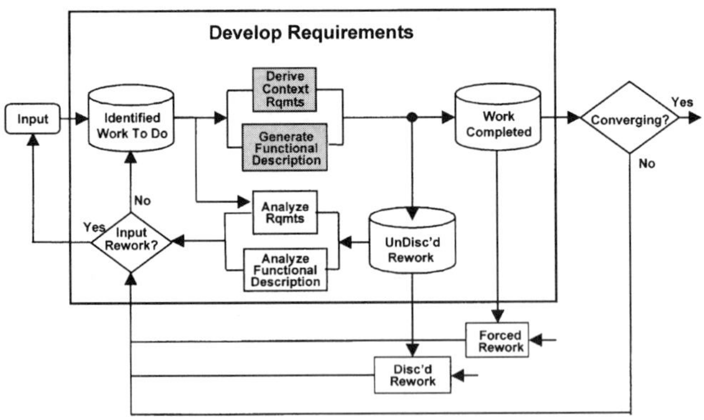

 26

INPUT to the Engineering Process

- Customer
  - Immediate
  - Procuring Organization
  - User Community
- Company
  - Corporation
  - Division
  - Department
- Previous Work
  - Business Development
  - Proposal
  - Previous Contract(s)/Studies
- Heritage Products/Designs
  - Division
  - Company
  - Competitor(s)
  - Customer(s)
- New Technologies
  - ITT IR&D, etc.
- Others

Rqmts Originate From Many Sources;
All Must Be Considered To Maximize Success

 27

"What" ➔ *Derive Context Rqmts*

- Determine context in which the system must function over its complete life-cycle
- Identify all mission phases, modes, and states
- Develop mission timeline & Operations Concept
- Identify all external interfaces by mission phase
- Identify critical issues by mission phase (events, technologies, etc.)
- Define environments by mission phase
- **Output**==>Prelim: Derived rqmts, Op's Concept, Context Diagram(s), Entity Relation Diagram(s), Event List(s), external ICDs, FMECA, etc.

 28

First, Identity "What" the System Must Do

Customer-Determined Design — **Functions**

Initial Customer Design — **Identify Major Functions**

- ⇨**Space Segment** → ⇨**Perform Satellite Operations**
  - ⇨**Telescope** → ⇨**Perform Telescope Operations**
  - ⇨**Spacecraft Bus** → ⇨**Support Telescope Operations**
- ⇨**Launch Segment** → ⇨**Perform Launcher Operations**
- ⇨**Ground Segment** → ⇨**Perform Ground Operations**

Required Functionality is Derived from the Design Concept

 29

Define Mission Phases

Parameters	**Mission Phases**				
	Integration & Test	**Deploy**	**Initialization**	**Operations**	**Disposal**
External Interfaces	Test Fixtures	Launcher Ground Sys	Launcher Ground Sys AKM	Ground Sys Relay Sats Other Sats	Ground System
Environment	Clean Room System Test	Air Ride Van Air Transport Launch site Facilities Fairing	Ascent Trajectory	Operational Orbit	Parking Orbit or Earth Re-Entry
System Modes	Test	Test Launch mode	On-Orbit test Maneuver Appendage Deploy	Nominal Standby Safe Maintenance On-Orbit test	De-Orbit

 30

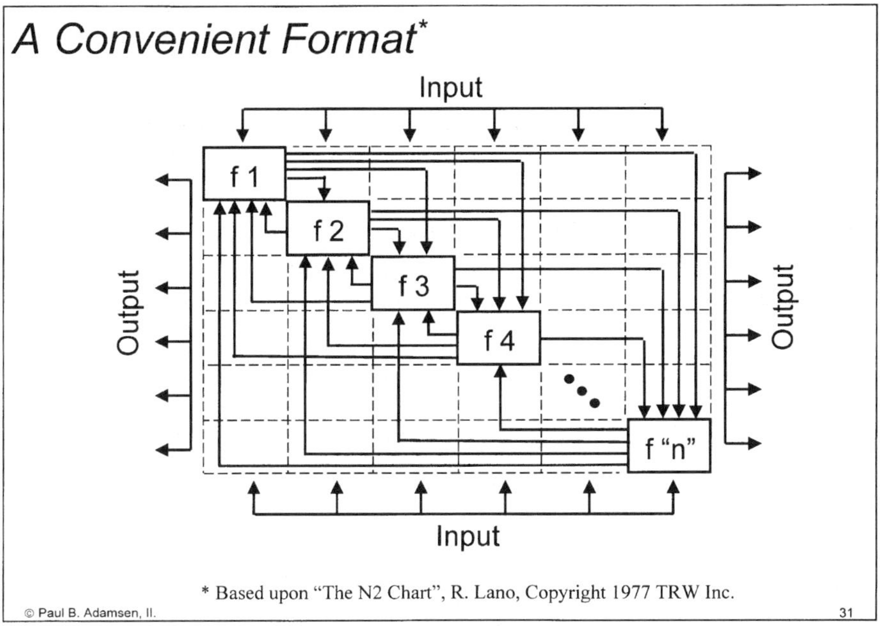
A Convenient Format*
Input
f 1
f 2
f 3
f 4
f "n"
Output
Output
Input
* Based upon "The N2 Chart", R. Lano, Copyright 1977 TRW Inc.
© Paul B. Adamsen, II.
31

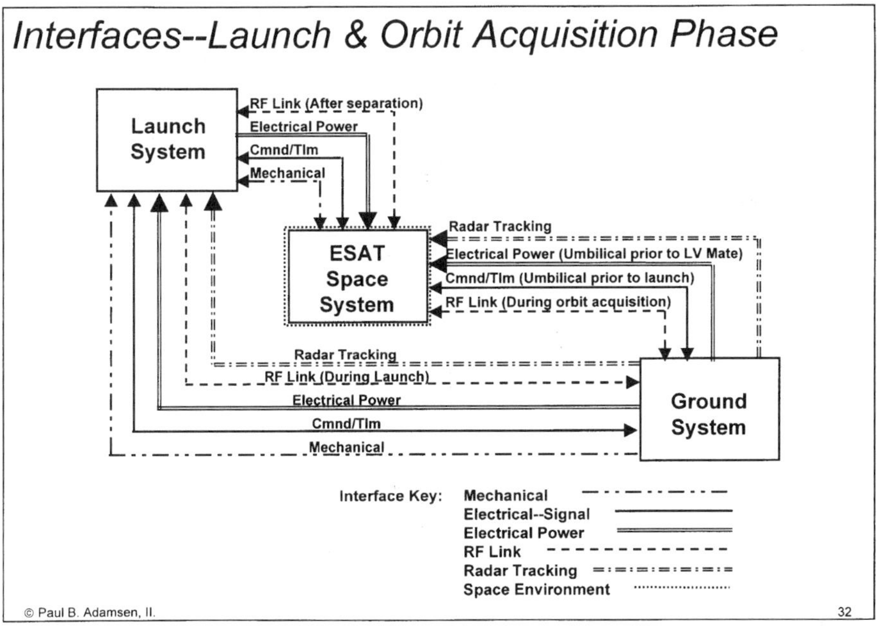
Interfaces--Launch & Orbit Acquisition Phase
Launch System
RF Link (After separation)
Electrical Power
Cmnd/Tlm
Mechanical
ESAT Space System
Radar Tracking
Electrical Power (Umbilical prior to LV Mate)
Cmnd/Tlm (Umbilical prior to launch)
RF Link (During orbit acquisition)
Radar Tracking
RF Link (During Launch)
Electrical Power
Cmnd/Tlm
Mechanical
Ground System
Interface Key:
Mechanical
Electrical--Signal
Electrical Power
RF Link
Radar Tracking
Space Environment
© Paul B. Adamsen, II.
32

"What" → Generate Funct'l Description

- Identify all functional requirements flowing out of imposed and derived requirements
- Identify performance requirements of each function and the relationships (interfaces, interdependencies, etc.) between functions
- Develop appropriate functional models
- Output → Validated spec(s), functional models (block diagrams, flow diagrams, behavior diagrams, etc.)
- Customer/Stakeholder Consensus

 33

Second, Develop Functional Block Diagrams-- Orbit Acquisition Phase

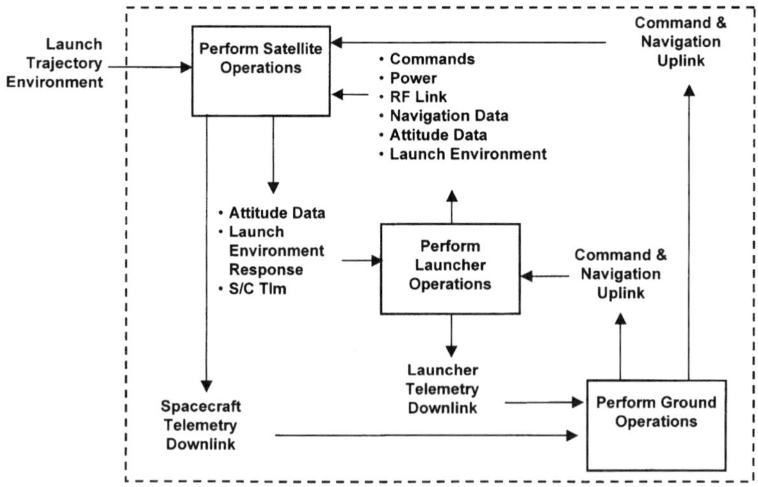

 34

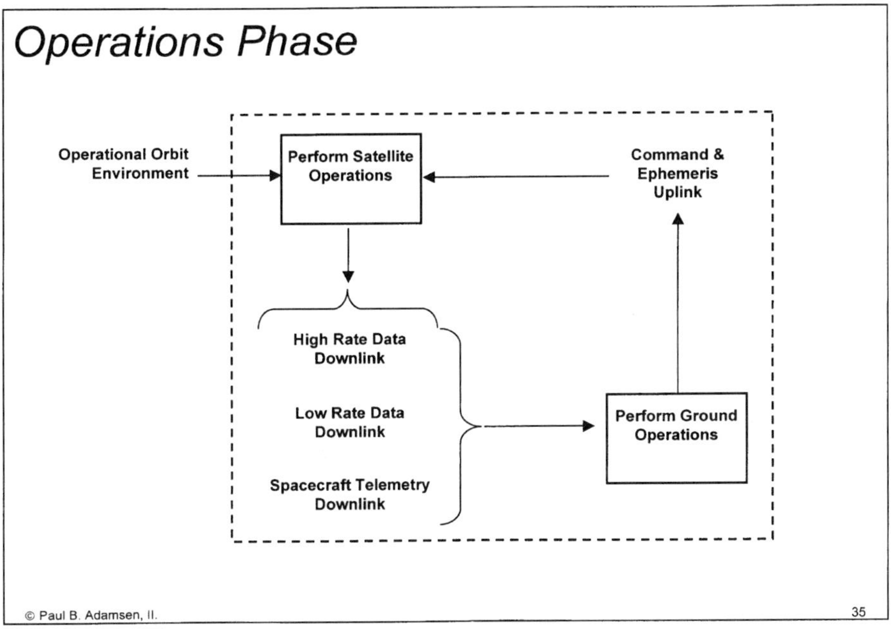
Operations Phase
Operational Orbit Environment
Perform Satellite Operations
Command & Ephemeris Uplink
High Rate Data Downlink
Low Rate Data Downlink
Spacecraft Telemetry Downlink
Perform Ground Operations
© Paul B. Adamsen, II.
35

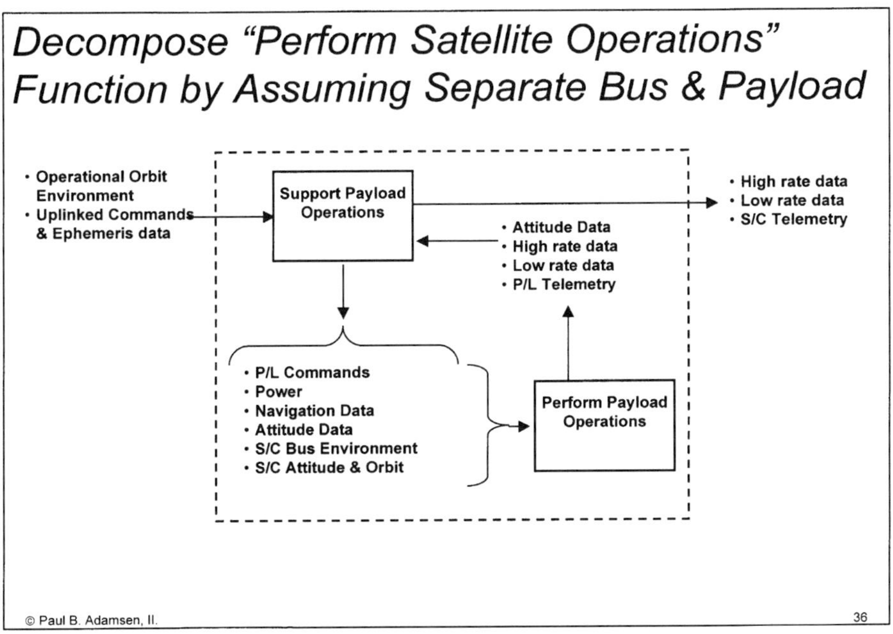
Decompose "Perform Satellite Operations" Function by Assuming Separate Bus & Payload
• Operational Orbit Environment
• Uplinked Commands & Ephemeris data
Support Payload Operations
• High rate data
• Low rate data
• S/C Telemetry
• Attitude Data
• High rate data
• Low rate data
• P/L Telemetry
• P/L Commands
• Power
• Navigation Data
• Attitude Data
• S/C Bus Environment
• S/C Attitude & Orbit
Perform Payload Operations
© Paul B. Adamsen, II.
36

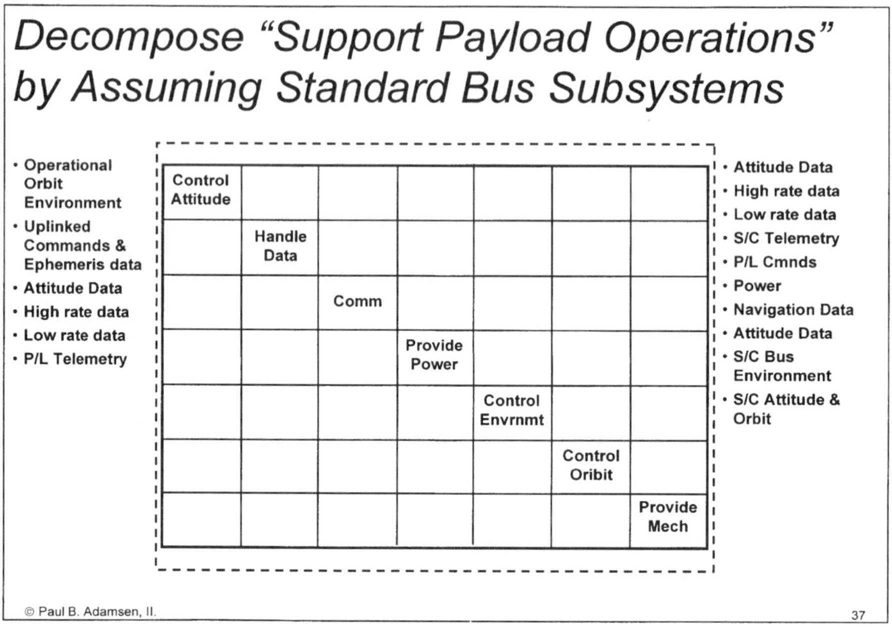
Decompose "Support Payload Operations" by Assuming Standard Bus Subsystems
• Operational Orbit Environment
• Uplinked Commands & Ephemeris data
• Attitude Data
• High rate data
• Low rate data
• P/L Telemetry
Control Attitude
Handle Data
Comm
Provide Power
Control Envrnmt
Control Oribit
Provide Mech
• Attitude Data
• High rate data
• Low rate data
• S/C Telemetry
• P/L Cmnds
• Power
• Navigation Data
• Attitude Data
• S/C Bus Environment
• S/C Attitude & Orbit
© Paul B. Adamsen, II.
37

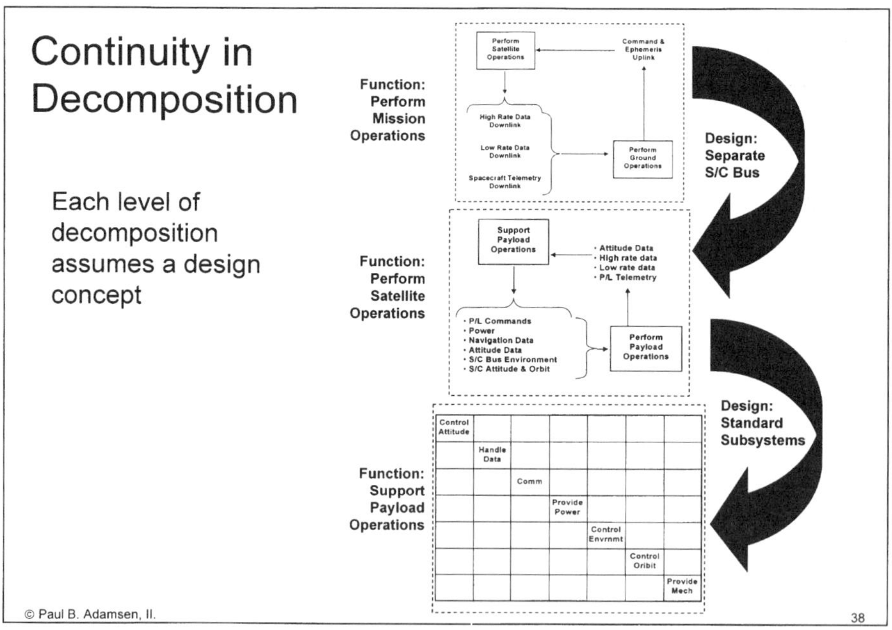
Continuity in Decomposition
Each level of decomposition assumes a design concept
Function: Perform Mission Operations
Perform Satellite Operations
Command & Ephemeris Uplink
High Rate Data Downlink
Low Rate Data Downlink
Spacecraft Telemetry Downlink
Perform Ground Operations
Design: Separate S/C Bus
Function: Perform Satellite Operations
Support Payload Operations
• Attitude Data
• High rate data
• Low rate data
• P/L Telemetry
• P/L Commands
• Power
• Navigation Data
• Attitude Data
• S/C Bus Environment
• S/C Attitude & Orbit
Perform Payload Operations
Design: Standard Subsystems
Function: Support Payload Operations
Control Attitude
Handle Data
Comm
Provide Power
Control Envrnmt
Control Oribit
Provide Mech
© Paul B. Adamsen, II.
38

Requirements Development: Rework Discovery Activities

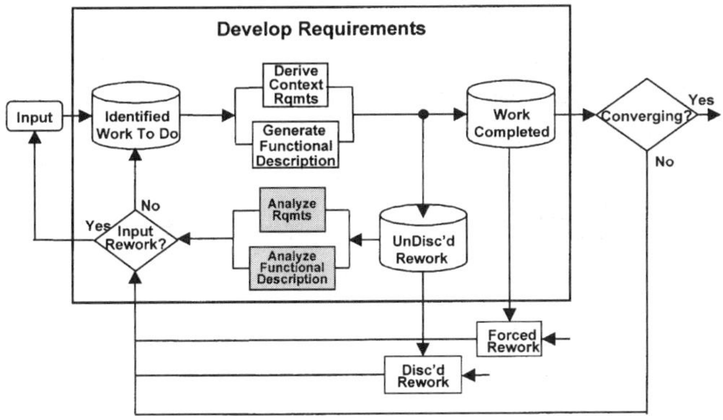

 39

"What" ➔ Analyze Requirements

- Identify all requirements, customer desires, customer priorities, constraints
- Analyze for completeness, consistency, etc.
- Interpret Customer rqmts & reach consensus
- Initialize rqmts database & maintain traceability

Output➔"Scrubbed" Requirements Set

 40

Analyze Rqmts Verifiability

- Determine if all imposed and derived requirements are verifiable
- Determine where in the system build-up each requirement will be verified
- Determine Verification method (Test, analysis, demonstration, simulation, inspection)
- **Output➔**Prelim: Verification Plan/Matrix, verifiable requirements statements

"What" ➔ *Analyze Funct'l Description*

- Verify adequacy of identified functions
  - Performance requirements
  - Output data sufficiency
- Verify interfaces between functions (timing, protocols, data content, etc.)
- Perform appropriate simulations, tests, etc.
- **Output** ➔ Validated spec(s), functional models (simulations, tests, etc.)

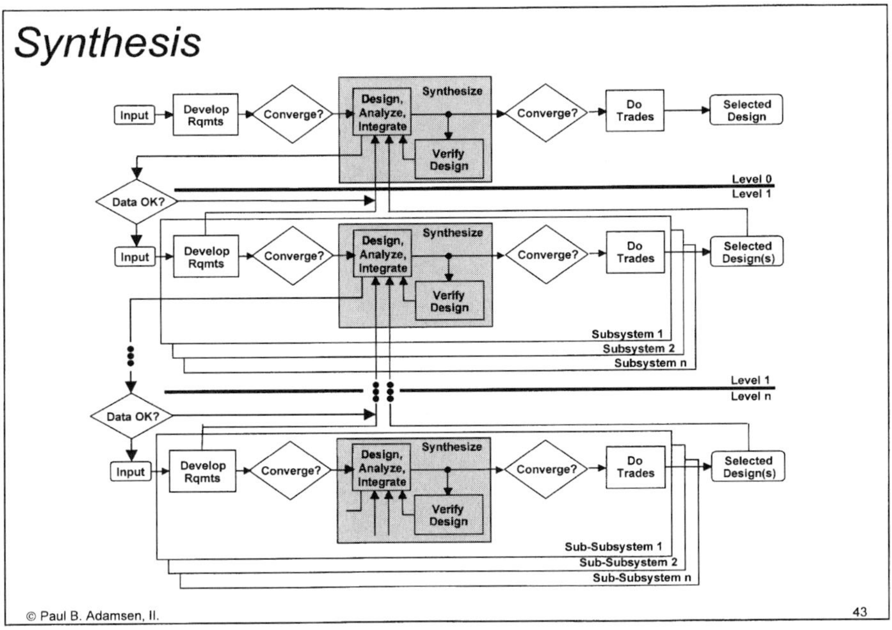
Synthesis
Input
Develop Rqmts
Converge?
Synthesize
Design, Analyze, Integrate
Verify Design
Converge?
Do Trades
Selected Design
Data OK?
Level 0
Level 1
Selected Design(s)
Subsystem 1
Subsystem 2
Subsystem n
Level 1
Level n
Sub-Subsystem 1
Sub-Subsystem 2
Sub-Subsystem n
© Paul B. Adamsen, II.
43

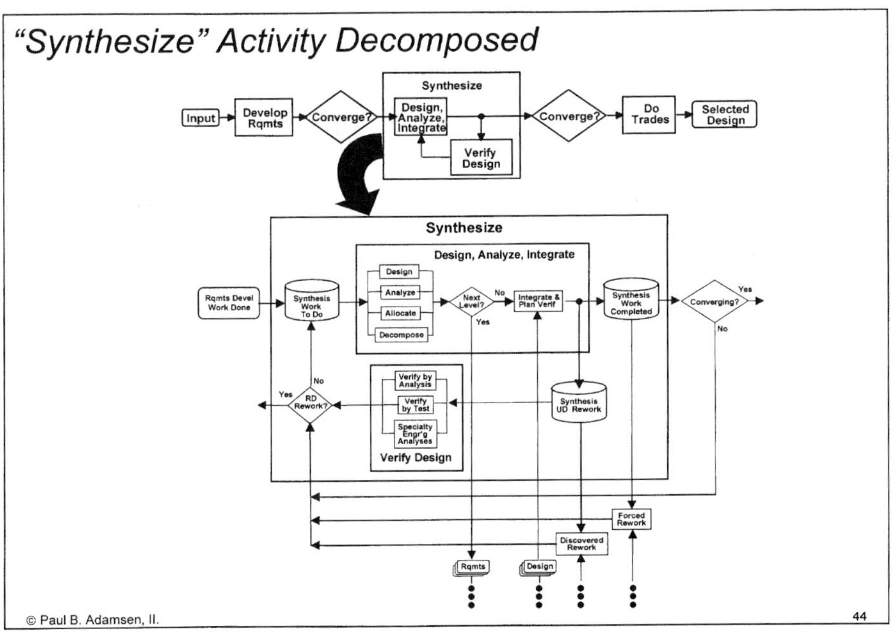
"Synthesize" Activity Decomposed
Input
Develop Rqmts
Converge?
Synthesize
Design, Analyze, Integrate
Verify Design
Converge?
Do Trades
Selected Design
Synthesize
Design, Analyze, Integrate
Rqmts Devel Work Done
Synthesis Work To Do
Design
Analyze
Allocate
Decompose
Next Level?
No
Yes
Integrate & Plan Verif
Synthesis Work Completed
Converging?
Yes
No
Verify by Analysis
Verify by Test
Specialty Engr'g Analyses
Verify Design
Synthesis UD Rework
RD Rework?
Forced Rework
Discovered Rework
Rqmts
Design
© Paul B. Adamsen, II.
44

Synthesis Work Activities —Design & Integrate

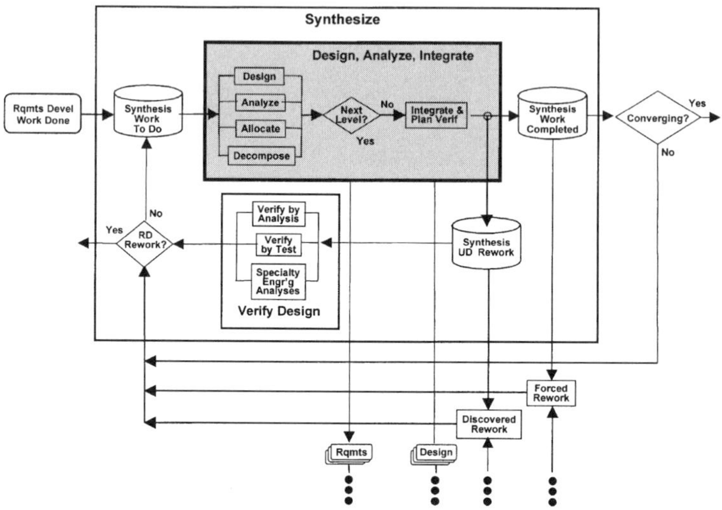

 45

Design

- All Necessary Disciplines Involved
  - –Engineering
  - –Manufacturing
  - –Integration & Test
  - –Operations, etc.
- Quantify Design Space (H/W & S/W)
  - –Parametric Analyses
  - –New Technologies
  - –Heritage Designs
- Generate Preliminary Design
  - – Deployed system and all support equipment
  - – Block diagrams, schematics, drawings, etc.
- Identify and Assess Risk
  - – Technical performance, cost, schedule
  - – Preliminary mitigation approaches
- **Output** ➔ Prelim: H/W & S/W design(s), risk assessment

 46

Existing S/C Configurations
Environmental Research Satellite
No Stabilization
Photo Courtesy TRW, Inc.
DSP
Spin Stabilization
Photo Courtesy TRW, Inc.
GEOSAT
Gravity Gradient
Photo Courtesy
JHU/APL
TDRSS
Bias Momentum
Photo Courtesy TRW, Inc.
Hubble Space Telescope
Zero Momentum
Photo Courtesy NASA
© Paul B. Adamsen, II.
47

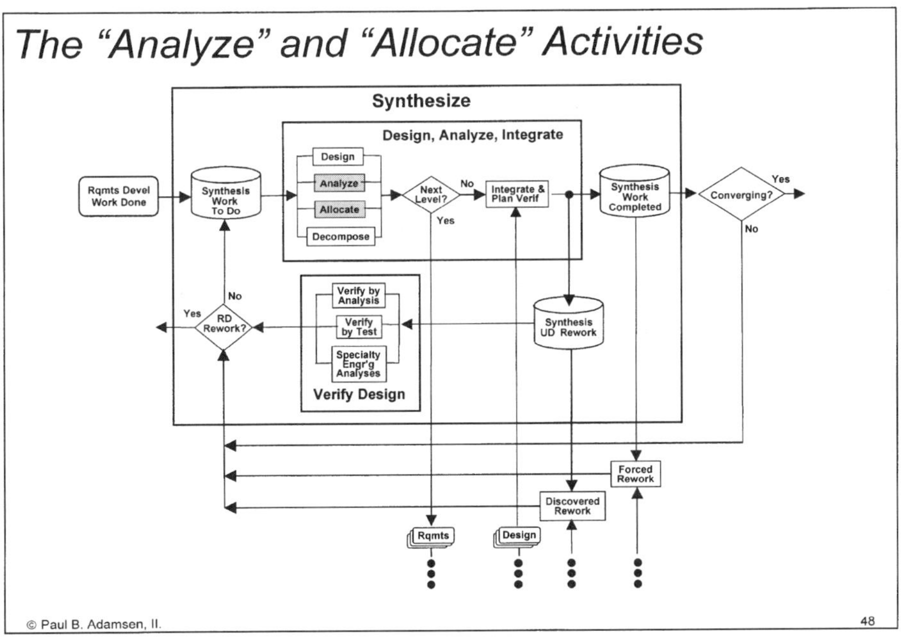
The "Analyze" and "Allocate" Activities
Synthesize
Design, Analyze, Integrate
Rqmts Devel Work Done
Synthesis Work To Do
Design
Analyze
Allocate
Decompose
Next Level?
No
Yes
Integrate & Plan Verif
Synthesis Work Completed
Converging?
Yes
No
Verify by Analysis
Verify by Test
Specialty Engr'g Analyses
Verify Design
RD Rework?
Yes
No
Synthesis UD Rework
Forced Rework
Discovered Rework
Rqmts
Design
© Paul B. Adamsen, II.
48

Allocation of Functionality to Implementation

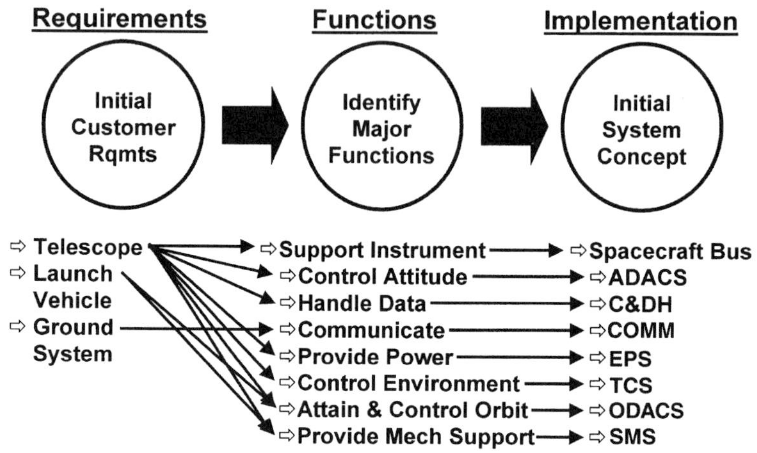

 49

Analysis

Any and all analyses necessary to quantify design space and parameters

- Mission--Context Definition
- Communications
- Command and Data Handling
- Electrical Power
- Environmental Control
- Propulsion
- Mechanical
- Attitude Determination and Control

 50

Allocation

- Allocate functionality, perf, constraints, etc. to system H/W and S/W elements
- Define preliminary budgets
  - Technical: mass, power, throughput, memory, RF links, etc.
  - Cost
  - Schedule
  - Programmatic: Risk, Reliability, Contamination, etc.
- Define Margin and Contingency Rules and Implement in Budgets
- Define/Refine TPMs
- Output ➔ Prelim budgets, TPMs

 51

Allocation ➔ Budgets

	Mass	Power	Memory	Thruput	Etc.
Communications					
Command and Data Handling					
Electrical Power					
Environmental Control					
Propulsion					
Attitude Determination and Control					

Margin Unknown Is Margin Lost!

 52

Notional Convergence of Margin and Reduction in Uncertainty

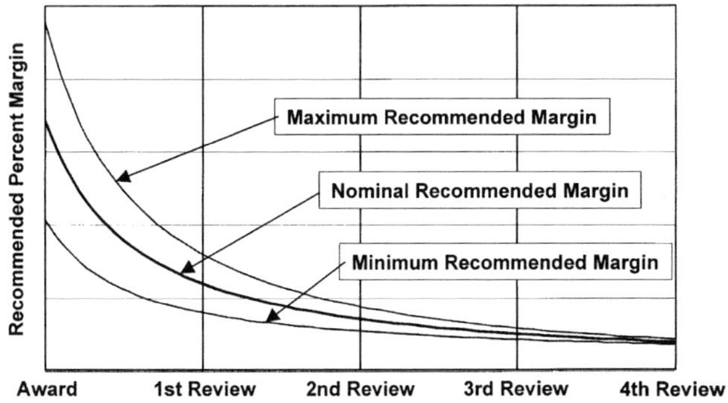

 53

Functional Decomposition Methodology

- Receive function and performance rqmts from RD activity
- Develop design concepts that meet rqmts
- Decompose concept into sub-functions for next-level down activity
- Identify interfaces between sub-functions
- Partition sub-functions into logical groups minimizing interfaces between groups
- Generate functional model and verify
- Generate function and performance rqmts (specs and ICD's) for each group
- Release function and performance rqmts to lower-level activities
- Receive feedback from lower-level development activities and refine specs and ICDs
- Iterate as necessary

Output → Lower level validated specifications, ICD(s), lower level element simulation

 54

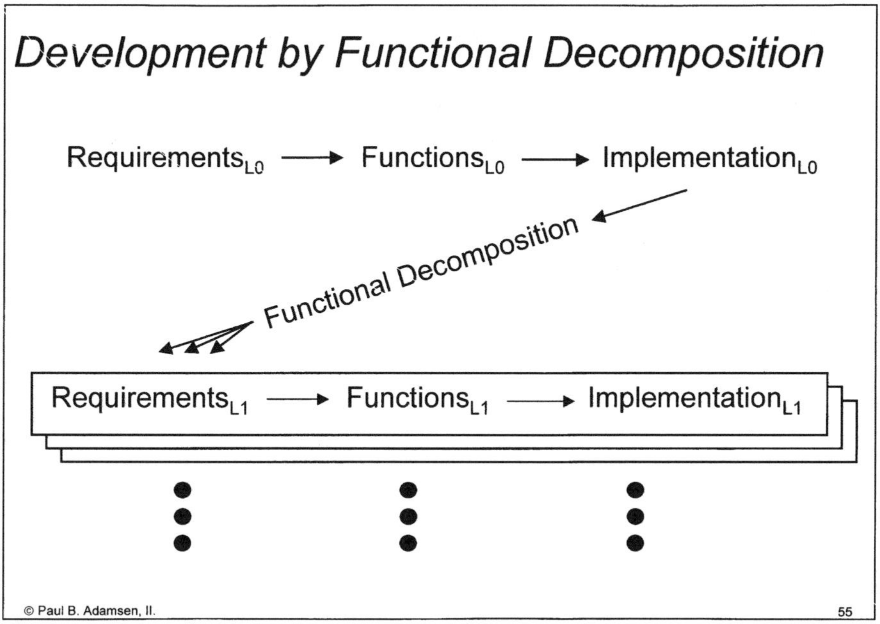
Development by Functional Decomposition
Requirements$_{L0}$ → Functions$_{L0}$ → Implementation$_{L0}$
Functional Decomposition
Requirements$_{L1}$ → Functions$_{L1}$ → Implementation$_{L1}$
© Paul B. Adamsen, II.
55

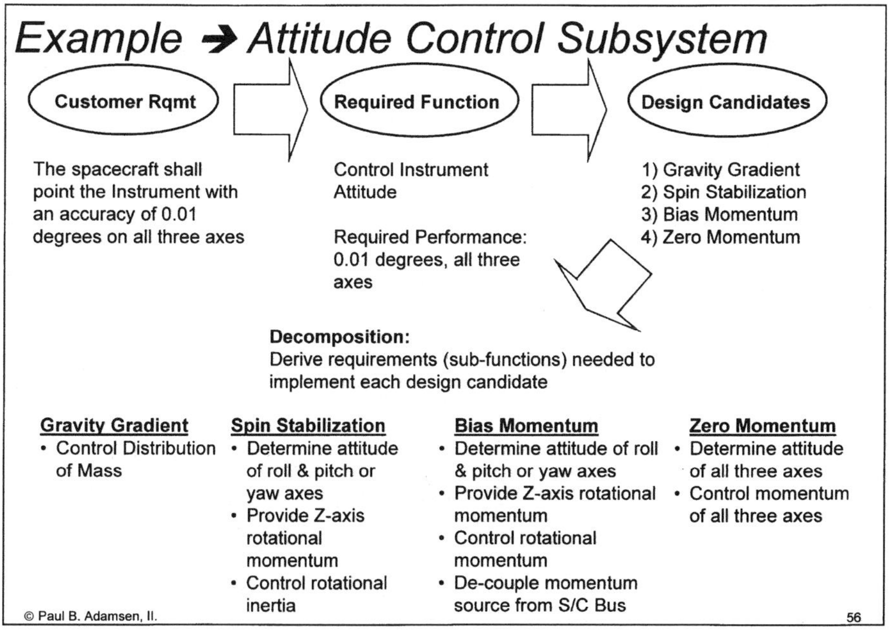
Example → Attitude Control Subsystem
Customer Rqmt
Required Function
Design Candidates
The spacecraft shall point the Instrument with an accuracy of 0.01 degrees on all three axes
Control Instrument Attitude
Required Performance: 0.01 degrees, all three axes
1) Gravity Gradient
2) Spin Stabilization
3) Bias Momentum
4) Zero Momentum
Decomposition:
Derive requirements (sub-functions) needed to implement each design candidate
Gravity Gradient
• Control Distribution of Mass
Spin Stabilization
• Determine attitude of roll & pitch or yaw axes
• Provide Z-axis rotational momentum
• Control rotational inertia
Bias Momentum
• Determine attitude of roll & pitch or yaw axes
• Provide Z-axis rotational momentum
• Control rotational momentum
• De-couple momentum source from S/C Bus
Zero Momentum
• Determine attitude of all three axes
• Control momentum of all three axes
© Paul B. Adamsen, II.
56

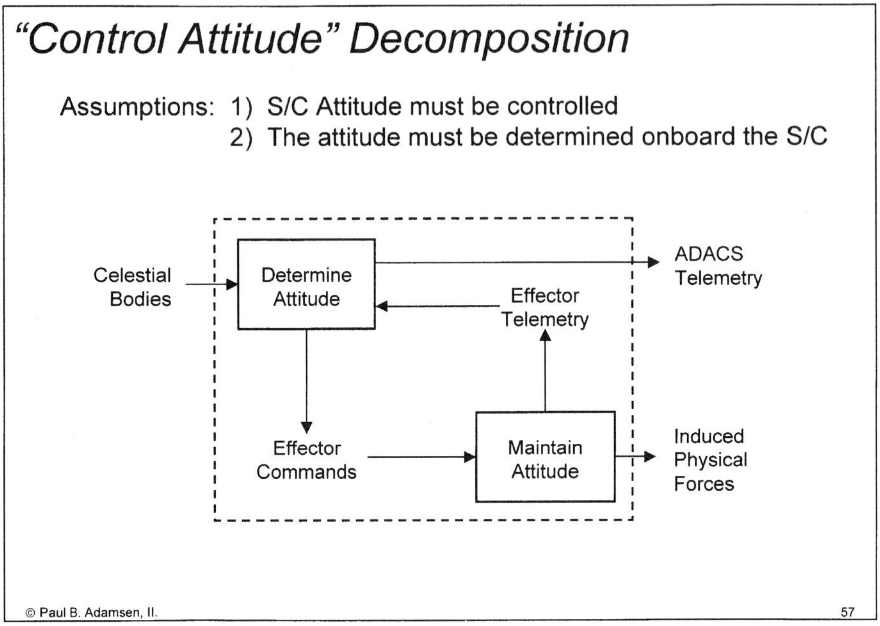
"Control Attitude" Decomposition
Assumptions: 1) S/C Attitude must be controlled
2) The attitude must be determined onboard the S/C
Celestial Bodies
Determine Attitude
ADACS Telemetry
Effector Telemetry
Effector Commands
Maintain Attitude
Induced Physical Forces
© Paul B. Adamsen, II.
57

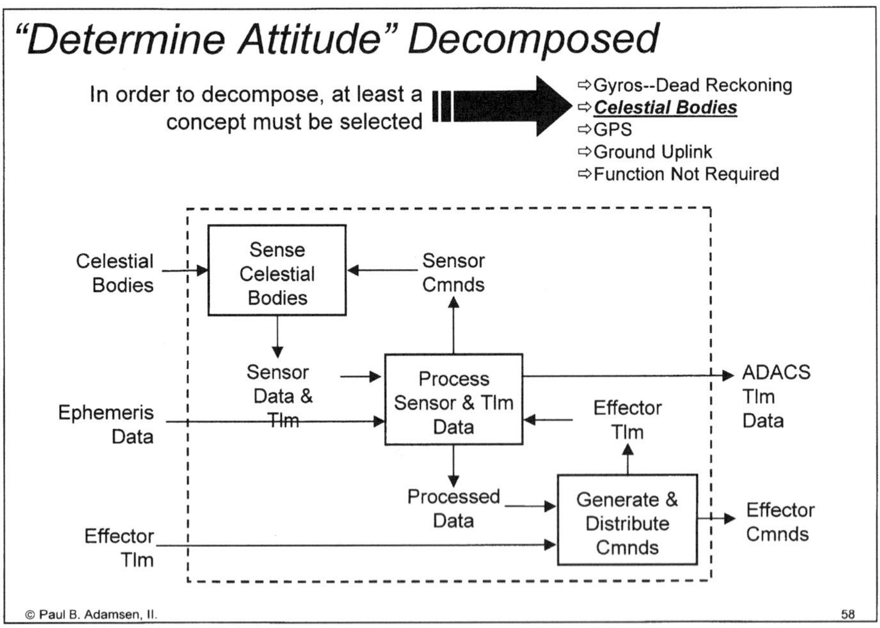
"Determine Attitude" Decomposed
In order to decompose, at least a concept must be selected
⇨Gyros--Dead Reckoning
⇨Celestial Bodies
⇨GPS
⇨Ground Uplink
⇨Function Not Required
Celestial Bodies
Sense Celestial Bodies
Sensor Cmnds
Sensor Data & Tlm
Process Sensor & Tlm Data
ADACS Tlm Data
Ephemeris Data
Effector Tlm
Processed Data
Generate & Distribute Cmnds
Effector Cmnds
Effector Tlm
© Paul B. Adamsen, II.
58

"Maintain Attitude" Decomposed

In order to decompose, at least a concept must be selected ➡

- ⇨Gyros--Dead Reckoning
- ⇨***Celestial Bodies***
- ⇨GPS
- ⇨Ground Uplink
- ⇨Function Not Required

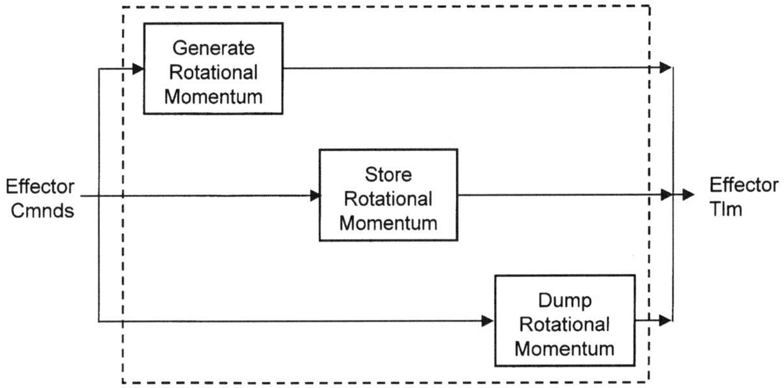

 59

Control Data Flow to Lower Level

- Configuration Control "Design-To" Information for Next-Tier Elements
- **Output** ➔ Configuration controlled documentation

 60

"Integrate & Plan Verification" Activity

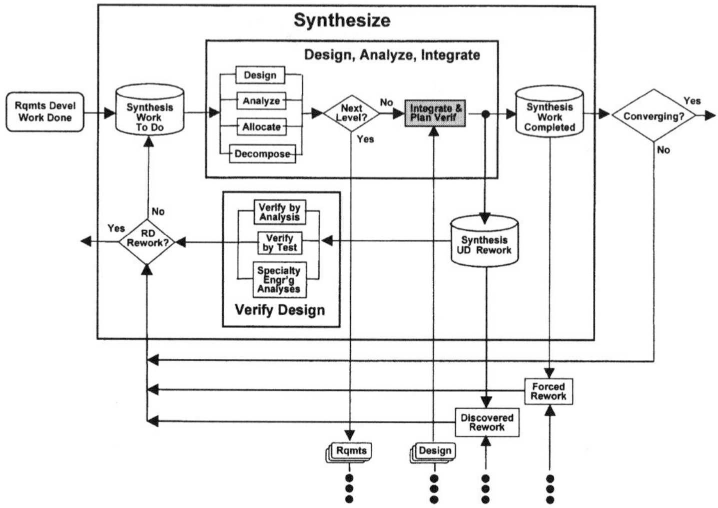

 61

Integrate

- Identify and Characterize I/Fs
- Control Configuration
  - Configuration Control Board
  - Handle Re-allocation
  - Specifications and ICDs
  - Databases, Etc.
- Update Mission Timeline/Operations Concept
- Update Design Definition
  - Block diagrams, schematics, Drawings, layouts
  - budgets, etc.
- Manage Risk
  - Update Risk Identification, Assessment, and Mitigation Approaches
- **Output** → Refined Design

 62

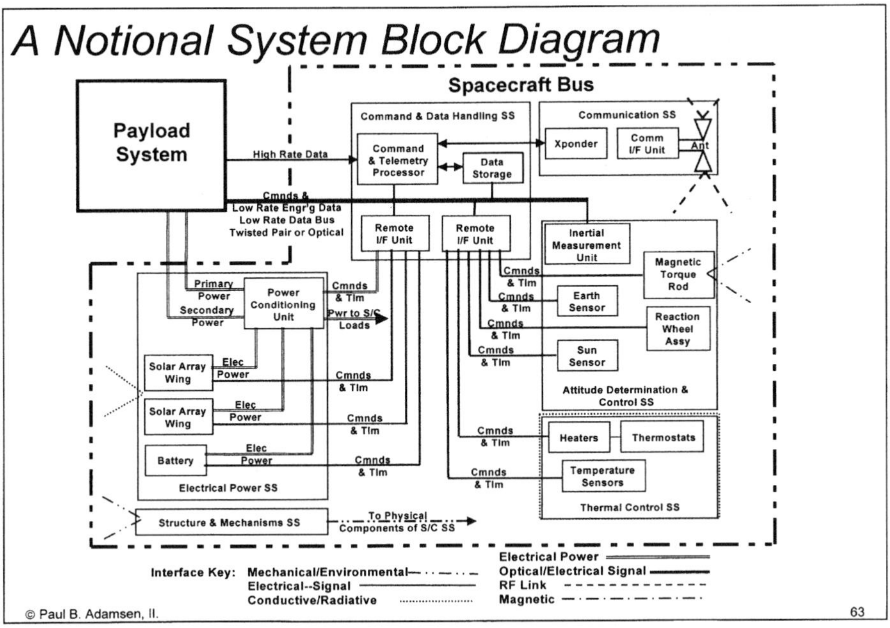
A Notional System Block Diagram
Spacecraft Bus
Payload System
High Rate Data
Cmnds & Low Rate Engr'g Data
Low Rate Data Bus
Twisted Pair or Optical
Command & Data Handling SS
Command & Telemetry Processor
Data Storage
Remote I/F Unit
Remote I/F Unit
Communication SS
Xponder
Comm I/F Unit
Ant
Inertial Measurement Unit
Magnetic Torque Rod
Earth Sensor
Reaction Wheel Assy
Sun Sensor
Attitude Determination & Control SS
Cmnds & Tlm
Primary Power
Secondary Power
Power Conditioning Unit
Pwr to S/C Loads
Solar Array Wing
Elec Power
Solar Array Wing
Battery
Electrical Power SS
Heaters
Thermostats
Temperature Sensors
Thermal Control SS
Structure & Mechanisms SS
To Physical Components of S/C SS
Interface Key:
Mechanical/Environmental
Electrical--Signal
Conductive/Radiative
Electrical Power
Optical/Electrical Signal
RF Link
Magnetic
© Paul B. Adamsen, II.
63

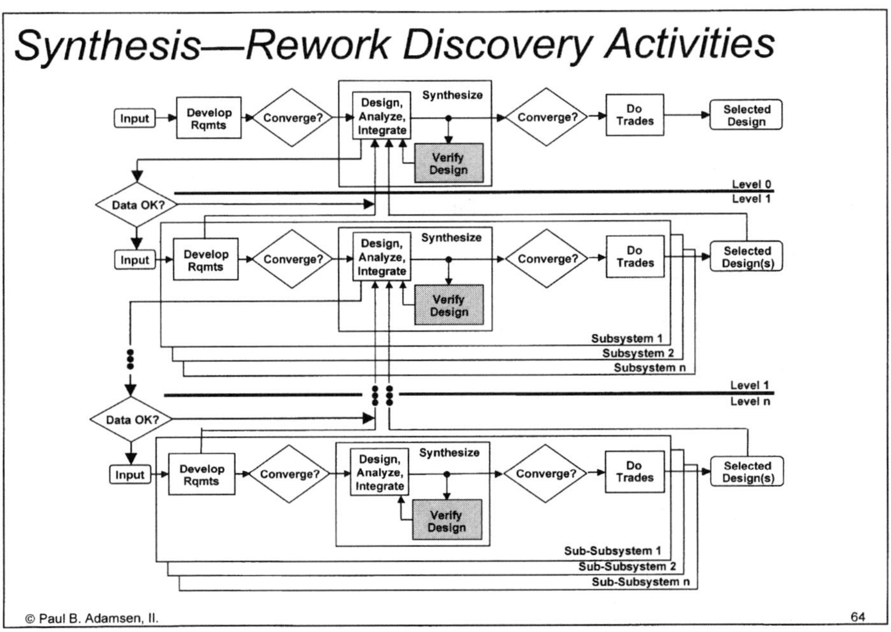
Synthesis—Rework Discovery Activities
Input
Develop Rqmts
Converge?
Synthesize
Design, Analyze, Integrate
Verify Design
Converge?
Do Trades
Selected Design
Data OK?
Level 0
Level 1
Selected Design(s)
Subsystem 1
Subsystem 2
Subsystem n
Level 1
Level n
Sub-Subsystem 1
Sub-Subsystem 2
Sub-Subsystem n
© Paul B. Adamsen, II.
64

Synthesis—Rework Discovery Activities

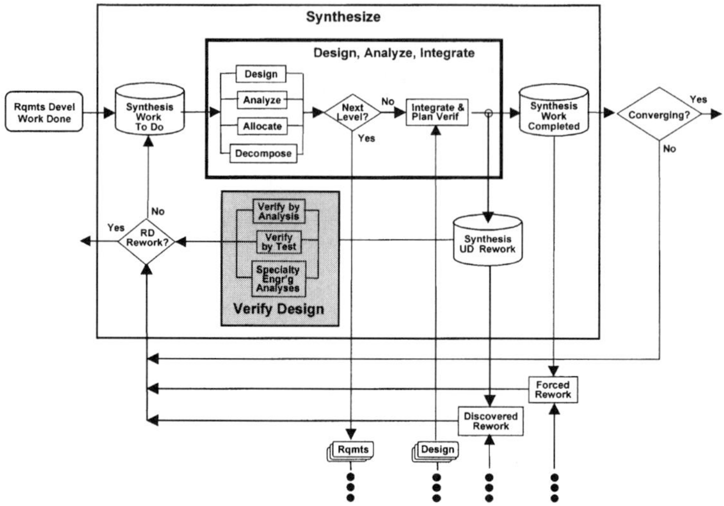

 65

Verify by Analysis

- Analysis—Analyses to determine "how well" the current design meets requirements
  - In contrast to analyses that define design space as described above in the design activity
- Output ➔ Mission, Electrical, and Mechanical analyses, simulations, etc.

 66

Verify by Test

Design Verification

- Pre-CDR, Integral part of design development activity
- Test Planning
  - –Test requirements
  - –Test flow
  - –Resource planning, Etc.
- Testing
  - –ETMs, prototypes, bread & brassboards, etc.
  - –System Verification Test (SVT)
- Output ➔ Verified Design

Product Verification

- Post CDR
- Focuses on the product to be deployed
- Test, analysis, simulation, demonstration, inspection
- Output ➔ Verified Product

Verification Activities Begin Early In The Program

 67

Specialty Engineering

- Is the design testable within resource and time constraints?
- Is the design producible within resource and time constraints?
- Is the design acceptable with respect to
  - –EMI/EMC
  - –Reliability
  - –Maintainability, affordability, supportability
  - –etc.

 68

Trade Analysis

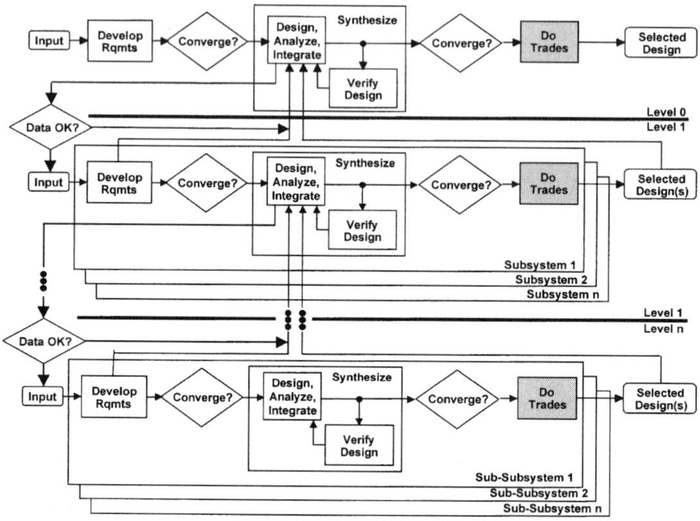

 69

Select

- Define Trade Criteria
  - Technical, Cost, Schedule, Risk
  - System Robustness
  - Sensitivity Analyses
- Assemble Trade Matrix
  - Must Have's, Wants, Utility
  - Value/Impact
- Assess Each Candidate
  - Select Best Design Values
- Selection
  - If multiple candidates are compliant and equally acceptable to design team, make selection at tier above
  - If selection is clear at the current tier, make selection
- Output ➔ Selected Design
- Rigor defined by program need

Trades Occur After Criteria Is Developed;
The Literature Provides Many Methodologies

 70

ADACS Candidate Technical Assessment

	Gravity Gradient	Spin Stabilization	Bias Momentum	Zero Momentum
Architecture	 GEOSAT *Photo Courtesy APL JHU*	 DSP Spacecraft *Photo Courtesy TRW*	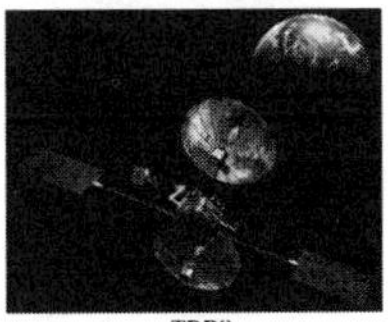 TDRS *Photo Courtesy TRW*	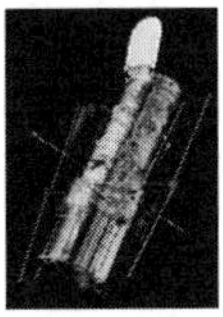 Hubble Space Telescope *Photo Courtesy NASA*
Design Elements	Passive Mass Dist	Aspect ratio Mass Balance Earth Sensor IMU, OBC Thrusters Mag Torqures	Earth Sensor Sun Sensor IMU, OBC Thrusters Momentum Wheel Mag Torquers	Earth Sensor Sun Sensor Star Tracker IMU, OBC Thrusters Reaction Wheels Mag Torquers
Accuracy	± 5° two axes	± 0.1° to ± 1°	± 0.1° to ± 1°	± 0.001° to ± 1°
Assessment	Cannot meet 0.01 accuracy rqmt	Spin not OK Cannot meet 0.01 accuracy rqmt	Cannot meet 0.01 accuracy rqmt	Meets all rqmts

 71

What About Optimization?

- It is not our purpose to discuss the myriad and specialized techniques
- It is our purpose is to describe where optimization occurs and how it impacts the overall process
- Optimization, by definition, implies a change to the design, therefore, the SDF provides feedback for it
- To some degree optimization occurs within each activity
- It is explicitly addressed here because it is at this point that technical, cost, and schedule criteria are available
- Elsewhere in the process, each design is being developed so optimization occurs through the iterations that naturally occur

 72

SDF 2nd Level Decomposition

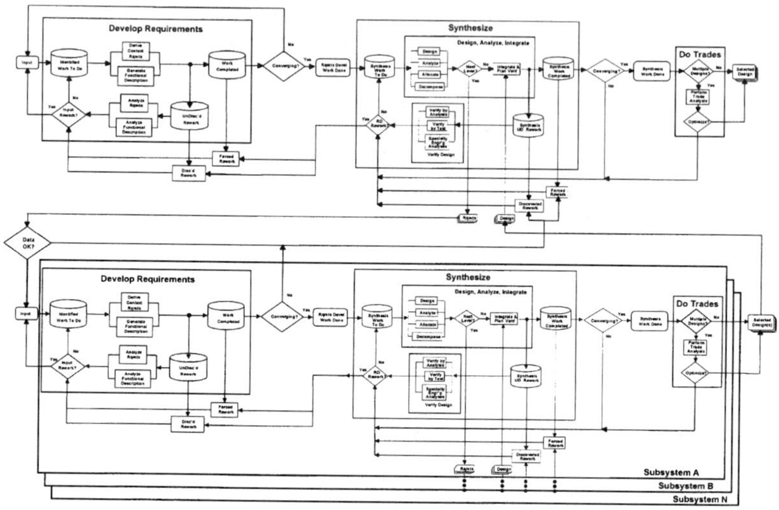

See page 90 for larger figure.

73

The SDF In The Time Domain

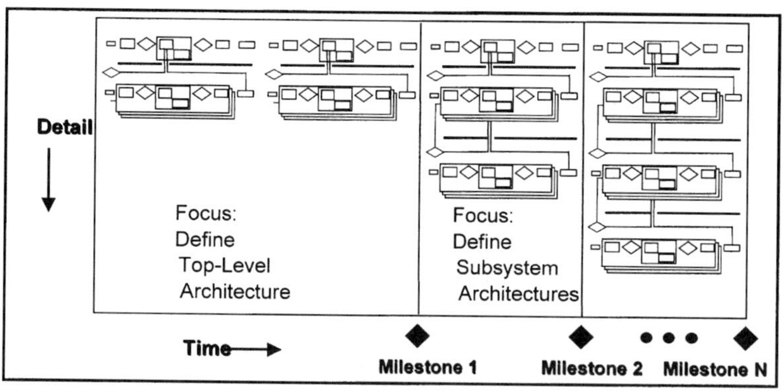

- One iteration of the EP may take "seconds" or much longer
- Incremental Solidification provides a mechanism for managing risk
- Outputs are defined as *f*(timeline)
- The SDF is iterated as required

74

Time-Phased Output at Planned Fidelity

System Tier	Award	First	Second	● ● ●	"N"
Level 0 Rqmts Level 0 Design	Draft Initial	Final Draft	Update Final	● ● ●	Update Update
Level 1 Rqmts Level 1 Design	Initial —	Draft Initial	Final Draft	● ● ●	Update Update
Level 2 Rqmts Level 2 Design	— —	Initial —	Draft Initial	● ● ●	Update Final
Level "n" Rqmts Level "n" Design	— —	— —	Initial —	● ● ●	Final Draft

 75

Full Life-Cycle

- Each mission phase imposes unique requirements on the system
- In order to maximize success, these requirements must be considered from the start
- The design effort generally continues up to the Critical Design Review
- After CDR, program moves from significant design effort to mfg, integration and test, deployment, op's, and disposal
- In production programs, provide feedback to the design activity capturing lessons learned from the deployed systems

 76

Full System Life-Cycle

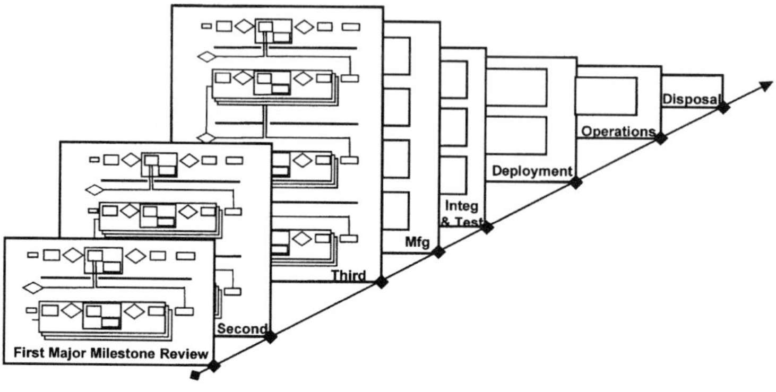

The Full System Life-Cycle must be Considered at the Earliest Stages of Development

77

Tailorability Of The SDF

- We have established that System Development Framework is the same for each level
- While not all SDF activities represent significant effort in every situation; generally, all are performed to some level of fidelity
- Therefore, tailoring is ***not*** done by changing the SDF, but by:
  - Effective partitioning of system elements (DSM)
  - Modulating the kinds and extent of documentation required
  - Modulating the level of detail and the scope of the activities performed

78

The System Development Framework:

Managerial

 79

Key Questions

- How should information flow and who is responsible for which interfaces?
- How should the Managerial aspects of system development be structured?
- How should the technical and managerial activities be coupled?

 80

What Are Mgmt & Technical Activities?

Managerial/Programmatic

- Configuration Management
- Risk Management
- Cost Management
- Schedule Management
- Customer Interface
- Subcontracts Management
- Etc.

Technical

- Requirements Development
- Synthesis
  - Design
  - Analysis
  - Integration
  - Verification
- Trades

These activities are closely coupled;
How they are coupled can be defined by the SDF!

 81

Program Structure and Control

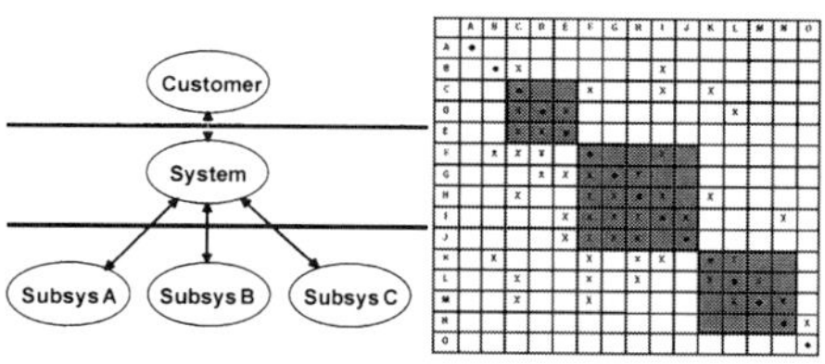

Start with Current Design, Partition elements such that the number of interfaces is minimized

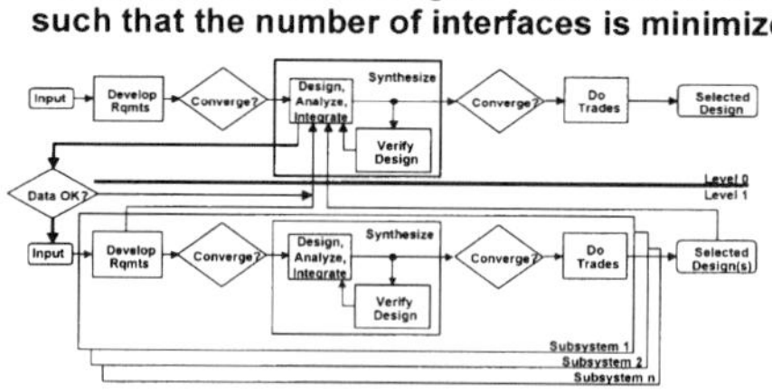

System Development Framework

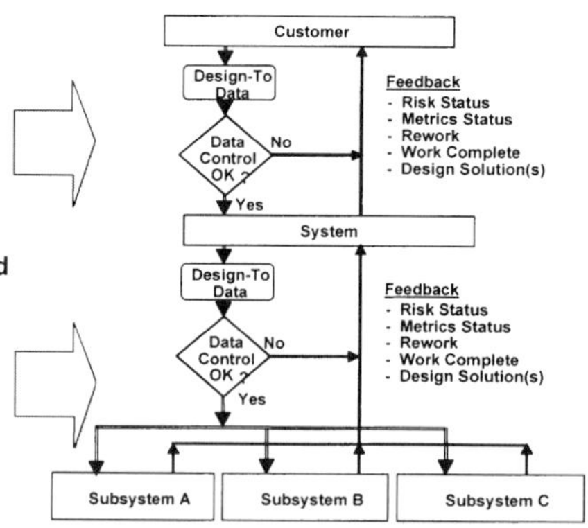

Resulting Team Structure & Control

Organize according to Architecture
Operate according to SDF Control Logic

 82

The SDF Defines IPT Interrelationships

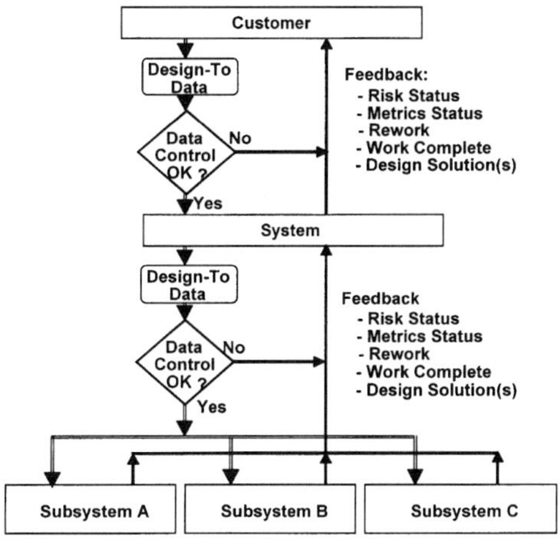

- **IPTs function independently within established bounds**
- **Configuration Management**
  - **Change Impact**
  - **Database Structure**
  - **TPMs, Budgets**
- **Risk Management**
  - **Periodic Review**
- **Roles & Responsiblities**
- **IPT Interfaces**
  - **ID, Characterization, Cntl**
- **Cost Management**

"Programmatic" I/Fs are defined by the SDF

 83

Exponential Growth in Complexity

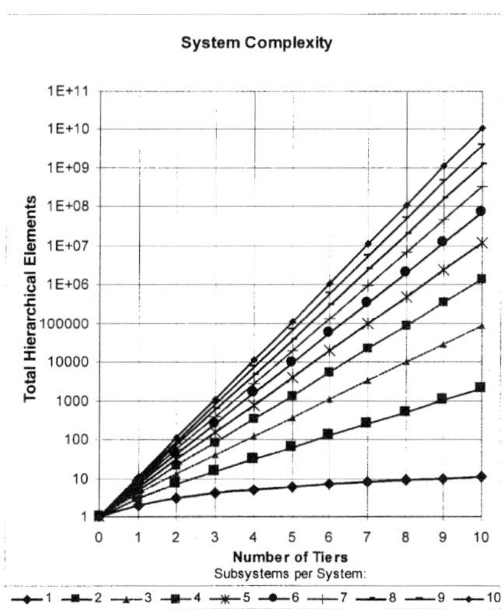

An Example: 3 subsystems per system, Find number of Total System Elements (TSE)

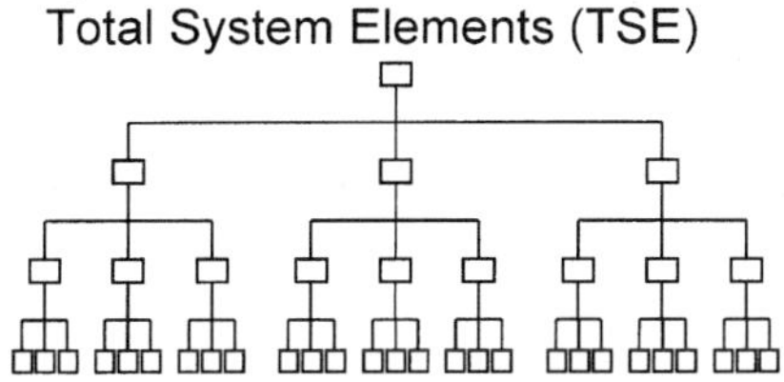

S = Number of Subsystems per System

m = Total number of Hierarchical levels

$$TSE = \sum_{n=0}^{m} S^n = 40$$

 84

Some SDF-Derived Principles

 85

Some Principles

- Acceptable risk is a key criterion in deciding to move from one activity to the next—The difficulty is accurately quantifying it
- Risk cannot be managed if it has not been identified and/or understood
- In conceptual architecting, the level of detail needed is defined by the confidence level desired
- A function cannot be decomposed. Only implementation can be decomposed. That which allows decomposition is knowledge about implementation

 86

More Principles

- Process understanding is no substitute for technical understanding—It is the technical understanding that enables development by decomposition
- Before a process can be improved it must be described
- Given our definition of "system", the same System Development Framework can be used at any tier of design development
- Costs due to rework increase exponentially with time

 87

Suggestions for Implementation in Industry—Focus on Output

- The SDF should be “Tailored” by identifying up front required inputs and outputs for each SDF activity
- Develop Exit Criteria for each review—These are derived directly from the outputs identified in the SDF
- Define required fidelity or “completeness” as a function of the program timeline.
- For each review, the program must produce the generalized SDF output—The structured approach is followed by default

 88

Bibliography

Adamsen, Paul B., Jr., A New Look at the System Engineering Process: A Detailed Algorithm, "Systems Engineering in the Global Market Place," *Proceedings of the Fifth Annual Symposium NCOSE,* Vol. 1, July 22-26, 1995, St. Louis, MO.

Adamsen, Paul B., Jr., Controlling The Chaos: An Integrated Approach To Managing A System Development Program, "Systems Engineering Practices and Tools," *Proceedings Sixth Annual Symposium INCOSE,* Vol. 1, July 7-11, 1996, Boston, MA, pp. 1093-1100.

Army Field Manual 770-78, System Engineering, Apr. 1979.

Blanchard, Benjamin S. and Wolter J. Fabrycky, *System Engineering and Analysis,* Englewood Cliffs, NJ: Prentice-Hall, 1981.

Brooks, Frederick P., Jr., *The Mythical Man-Month: Essays on Software Engineering, Anniversary Edition,* Reading, MA: Addison-Wesley Longman, 1995.

Chase, Wilton P., *Management of System Engineering,* New York: John Wiley & Sons, 1974.

Chestnut, Harold, *System Engineering Tools,* New York: John Wiley & Sons, 1965.

Clayson, Mark H., Heuristic Structuring and Methods Selection in the SEP: An Expert System in M.4, "Systems Engineering Practices and Tools," *Proceedings Sixth Annual Symposium INCOSE,* Vol. 1, July 7-11, 1996, Boston, MA, pp. 933-942.

Cooper, Kenneth G., Naval Ship Production: A Claim Settled and a Framework Built, *Interfaces,* 10:6, Dec. 1980, The Institute of Management Sciences.

Cooper, Kenneth G. and Thomas W. Mullen, Swords and Plowshares: The Rework Cycles of Defense and Commercial Software Development Projects, *American Programmer,* May 1993.

de S. Coutinho, John, *Advanced Systems Development Management,* New York: John Wiley & Sons, 1977.

Defense Systems Management College (DSMC), *Systems Engineering Management Guide,* Technical Management Department, Dec. 1986.

Defense Systems Management College (DSMC), *Systems Engineering Management Guide,* Technical Management Department, Jan. 1990.

Dommasch, Daniel O. and Charles W. Laudeman, *Principles Underlying Systems Engineering,* New York: Pitman Publishing, 1962.

Electronic Industries Association (EIA), EIA Interim Standard Systems Engineering, EIA/IS 632, Engineering Department, Washington DC, 1994.

Ellis, David O. and Fred J. Ludwig, *Systems Philosophy,* Englewood Cliffs, NJ: Prentice-Hall, 1962.

Eppinger, Steven D., Daniel E. Whitney, Robert P. Smith, and David A. Gebala, A Model-Based Method for Organizing Tasks in Product Development, in *Research in Engineering Design*, 6:1-13, 1994.

Ford, David N., The Dynamics of Project Management: An Investigation of the Impacts of Project Process and Coordination on Performance, MIT Doctoral Dissertation, 1995.

Ford, David N. and John D. Sterman, Dynamic Modeling of Product Development Processes, Working Paper, MIT Sloan School of Management, Jan. 1997.

Forrester, Jay, *Industrial Dynamics*, Portland, OR: Productivity Press, 1961.

Gulati, Rosaline K. and Steven D. Eppinger, The Coupling of Product Architecture and Organizational Structure Decisions, MIT Sloan School of Management, International Center for Research on The Management of Technology, Working Paper Number 151-96.

Hall, Arthur D., *A Methodology For Systems Engineering*, Princeton, NJ: D. Van Nostrand, 1962.

Hatley, Derek J. and Imtiaz A. Pirbhai, *Strategies For Real-Time System Specification*, New York: Dorset House, 1988.

Institute of Electrical and Electronics Engineers (IEEE), *IEEE Trial-Use Standard for Application and Management of the Systems Engineering Process*, IEEE 1220-1994, New York, 1995.

Kraniauskas, Peter, *Transforms in Signals and Systems*, Wokingham, England: Addison-Wesley, 1993.

Krishnan, Viswanathan, Steven D. Eppinger, and Daniel E. Whitney, A Model-Based Framework to Overlap Product Development Activities, *Management Science*, Vol. 43, No. 4, pp. 437-451, Apr. 1997.

Larson, Wiley J. and James R. Wertz, *Space Mission Analysis and Design*, Kluwer Academic, Dordrecht, The Netherlands, 1992.

McCord, Kent R. and Steven D. Eppinger, Managing the Integration Problem in Concurrent Engineering, MIT Sloan School of Management, Working Paper Number 3594, Aug. 1993.

McCumber, William H., System Performance Representation: Standard Scoring Functions, NCOSE 1995, P003.

Merriam-Webster's Collegiate Dictionary, Tenth Edition, Springfield, MA: Merriam-Webster, 1996.

MIL-STD-499A, Engineering Management (USAF).

MIL-STD-499B.

Minds, Kevin S., System Engineering The People System, NCOSE 1995, P066.

Pimmler, Thomas U. and Steven D. Eppinger, Integration Analysis of Product Decompositions, *Design Theory and Methodology*, DE-Vol 68, ASME 1994.

Pugh, Stuart, *Total Design*, Wokingham, England: Addison-Wesley, 1992.

Rechtin, Eberhardt, *Systems Architecting: Creating and Building Complex Systems*, Englewood Cliffs, NJ: Prentice-Hall, 1991.

Rechtin, Eberhardt and Mark W. Maier, *The Art of Systems Architecting*, Boca Raton, FL: CRC Press, 1997.

Reinert, Richard P. and James R. Wertz, *Space Mission Analysis and Design*, Eds. Wiley J. Larson and James R. Wertz, published jointly by Microcosm, Inc., Torrance, CA and Kluwer Academic Publishers, The Netherlands, 1992.

Richardson, George P. and Alexander L. Pugh, *Introduction to System Dynamics Modeling*, Portland, OR: Productivity Press, 1981.

Rochecouste, Hervé, A Systems Engineering Capability in the Global Market Place, NCOSE 1995, P004.

Shinners, Stanley M., *A Guide to Systems Engineering and Management*, Lexington, MA: D. C. Heath and Company, 1976.

Smith, Robert P. and Steven D. Eppinger, Identifying Controlling Features of Engineering Design Iteration, *Management Science*, Vol. 43, No. 3, pp. 276-293, March 1997.

SPC, A Tailorable Process for Systems Engineering, SPC-94095-CMC, Version 01.00.05, 1995.

Suh, Nam P., *The Principles of Design*, New York: Oxford University Press, 1990.

Ulrich, Karl T. and Steven D. Eppinger, *Product Design and Development*, New York: McGraw-Hill, 1995.

Wertz, James R., Ed., *Spacecraft Attitude Determination and Control*, D. Reidel, Dordrecht, The Netherlands, 1997 reprint.

White, Michelle M., James A. Lacy, and Edgar A. O'Hair, Refinement of the Requirements Definition (RD) Concept in a System Development: Development of the RD Areas, "Systems Engineering Practices and Tools," *Proceedings Sixth Annual Symposium INCOSE*, Vol. 1, July 7-11, 1996, Boston, MA, pp. 749-756.

Wymore, Wayne A., *Model-Based Systems Engineering*, Boca Raton, FL: CRC Press, 1993.

Wymore, Wayne A., *A Mathematical Theory of Systems Engineering — The Elements*, New York: John Wiley and Sons, 1967.

Index

A

B

C

D

S

Z